THE FUNDAMENTALS OF

LEISURE OFF-ROAD RIDING

ON TWO WHEELS

ISBN: 978-989-33-6093-4

INDEX

INTRO..1

CHAPTER I - WHAT'S ADVENTURE OFF-ROAD RIDING ALL ABOUT..6

TERRAIN RATINGS..8

TRAIL ETIQUETTE AND SAFETY..10

HYDRATION..11

DON'T GET RUN OVER..14

FIRST AID KIT..16

SHOW THE RESPECT YOU WANT TO RECEIVE..19

CHAPTER II - THE RIGHT BIKE..22

FROM THE SEAT TO THE GROUND..24

FROM THE SEAT TO THE PEGS..27

FROM PEGS TO FOOT LEVERS..28

FROM THE HANDLEBAR TO THE LEVERS..28

FROM THE TORSO TO THE HANDLEBAR..31

WEIGHT..32

POWER..34

CHAPTER III - TRACTION AND GEOMETRY..37

KAMM CIRCLES..38

HYSTERESIS..46

TIRE DYNAMICS IN A TURN..47

TRUST YOURSELF..52

LET THE FRONT END WORK..53

THE WHEEL ROLLS OVER..56

CHAPTER IV - THE MIND OF THE RIDER 60
USING EYESIGHT WITH INTENT 60
THE MICRO ADJUSTMENT PROBLEM 65
BREATHING CONTROL 68

CHAPTER V - KNOW YOUR BIKE 75
GENERAL MEASUREMENTS 79
TOO MUCH OF A GOOD THING 80
EVERY BIKE HAS TWO SIDES 82

CHAPTER VI - BODY POSITIONING FROM HEAD TO TOE 87
FEET 93
FROM THE FEET TO THE KNEES 107
HIP AND TORSO 110
ARMS 117
HANDS 121
LATERAL AND LONGITUDINAL MOVEMENTS 126
THE LIMITS OF THE BIKE AND THE BODY 132
BASIC DYNAMIC MOVEMENTS 133
UPHILLS 137
DOWNHILLS 140
EMERGENCY BRAKING 145
TURNS 148

CHAPTER VII - PERCEPTION MISTAKES 156
OFF-ROAD IS ALSO RIDING SITTING DOWN 156
THE TWO SEATED POSITIONS 159
THE TWO STANDING POSITIONS 161
THE ELUSIVE 5TH POSITION 163
LIFTING THE BIKE OFF THE GROUND 165
WHEN IN DOUBT, BRAAPP IT OUT 167

TRAINING ISN'T EXPENSIVE 171

CHAPTER VIII - THE VALUE OF IGNORANCE 177

ANNEX I - RIDER AND TERRAIN CLASSIFICATION SYSTEM 182

CONSIDERANTIONS AND ASSUMPTIONS 183

ADVENTURER CLASSIFICATION SYSTEM 186

TERRAIN CLASSES 188

CLASS 1 / BEGINNER TERRAIN ⊖ 188

CLASS 2 / BASIC TERRAIN ● 188

CLASS 3 / MODERATE TERRAIN ■ 190

CLASS 4 / HARD TERRAIN ◆ 191

CLASS 5 / EXTREME TERRAIN ◆◆ 192

ANNEX II - FIRST AID CHEAT SHEET 194

REFERENCES 195

Charlotte, Benny, and Olivia, thank you for your patience, love, and trust in both me and my work. Without your support and understanding, what I do would truly be impossible.

Thank you!

Dave, you're a legend, thanks!

INTRO

Over a century since the first motorcycle was invented, there's no denying the technological evolution of these machines. But have our skills as riders kept pace with these advancements? Broadly speaking, the answer is a resounding no. And with the rise of electronic aids that increasingly allow the bike to "ride for us" or at least play a crucial part in the riding process, there's little to suggest any organic shifts in rider development.

This isn't to say there's been a direct increase in rider safety. If we draw parallels to aviation, renowned for its safety and consistency, despite today's aircraft being capable of flying from takeoff to landing on autopilot, pilots still require rigorous and ongoing training to maintain proficiency.

That being said, instead of resigning ourselves to the technological takeover of our bikes, I believe that we, as a riding community, should be delving into how our bikes work, how to master their operation with and without electronic aids, and to grasp every facet that defines our sport, far beyond mere riding techniques.

A motorcycle isn't a car or a plane, with their myriad of isolated or backup safety mechanisms designed to protect us. Even the most advanced bike can't compensate for a puncture, excessive body movements, adverse weather, or a basic misunderstanding of our sport's unwritten rules – all factors that delineate the line where the bike's control ends and the rider's reaction begins.

Drawing lessons from aviation safety statistics, embracing the technical aspects of motorcycling will significantly enhance our safety, often making the difference between life and death. Regardless of riding pace – be it that of a leisurely weekend warrior or a competitive athlete – knowledge breeds consistency, which in turn reduces mistakes.

Imagine facing a turn every 300 feet over a 60-mile ride; that's a little over 1,000 turns in one single day. Attacking each turn with consistent technique not only bolsters safety but also inadvertently boosts our cruising speed, allowing for a more comfortable and exciting ride with ample margin for errors.

Thus, technique transcends mere physical postures like "elbow up" or "ass back." It's a compendium of skills that, when combined, empower riders to analyze situations and make optimal choices, divided into understanding off-roading, bike mechanics and physics, bodily awareness, control mastery, and balanced positioning.

With a foundational grasp of these aspects, even at a basic level, we transition from novices to experienced riders, setting the stage for a journey toward becoming the secure, consistent riders we should all aim to be. With that in mind, what follows in this book isn't solely my personal revelations but a communal learning distilled from diverse fields, from neuroscience to physics, underpinning the collective wisdom that shapes our riding ethos.

As humans, our cognitive evolution is exponential because, unlike other species, each generation develops from the knowledge of all that came before. The engines powering our motorcycles today, for example, are the culmination of nearly 200 years of development, with each innovation built on the theoretical and practical knowledge of predecessors. Unfortunately, in our off-road two-wheeled world, we, as recreational riders, tend to learn in isolation, or from others who, having chosen the same empirical path, may only be a few steps ahead.

This approach is problematic in a sport where many techniques are counterintuitive, and it's also why we often see little change in the technical mistakes made by riders across generations. Thankfully, an increasing number of riders are beginning to shift from this past trend, focusing on foundational training and sharing ideas and concepts beyond basic riding techniques, such as understanding the differences amongst various off-road terrains.

This change, while subtle, points to a future where the contents of this book will be, hopefully and rightfully, outdated or incomplete. With an awareness that this shift should be backed by continuous education and information, this book serves then as a compilation of what I've learned, observed, and gathered from students, instructors, professionals from various fields, and athletes over my last decade as a professional instructor.

As part of my contribution to the ongoing evolution of our sport, this book emphasizes the importance of seeking well-grounded practical and theoretical information and encourages us to question the "why" behind what we're told, instead of just blandly accepting the "how." These are the foundations that will enable us to get the most out of ourselves, our bikes, and our sport.

However, it is still important to clarify that this book is not a direct "how-to" to be followed blindly, nor is it a scientific paper free from errors, omissions, or shortcuts. It should be seen as a guide from A to Z, aiming to open new horizons, challenge preconceived notions, and bring to the forefront topics that are often overlooked or undervalued in our field.

While many chapters delve into detail, some concepts and details are presented in a more cursory manner, as the goal is to open the door to the main point: understanding the basics of why certain techniques are used. This approach, which I regularly employ in training sessions, yields excellent results by providing students with a comprehensive understanding of the "why" behind each technical movement, a more critical overall view of technique, and a solid foundation for personal exploration and deeper learning.

By becoming more complete riders mentally, we can define boundaries, understand forces, and expand our comfort zones, allowing us to be more consistent, safe, and continuously evolving.

CHAPTER 1

WHAT'S ADVENTURE OFF-ROAD RIDING ALL ABOUT

[illegible]

[illegible]

[illegible]

CHAPTER I

WHAT'S ADVENTURE OFF-ROAD RIDING ALL ABOUT

I'd say most of us who love off-road motorcycles see our sport as super inclusive. You can tear it up on big or small bikes, with two-strokes or four, running on gas or going electric. It's something you can do from age 8 to 80, and it doesn't matter who you are or what you believe in.

But, I reckon there are a few spots where we, as a community, could do better, and topping that list is getting clear on the fact that there are all kinds of "dirt" out there. Even though I believe the off-road riding community is tight-knit, it's pretty common to scroll through an online post about a ride or overhear at a coffee shop someone quipping, "That ain't real off-road," or something along those lines.

Now, it's not my mission in this book to unpack why that sort of talk isn't socially acceptable, but I do want to shed light on why those oversimplified comments are still rooted in an understandable base.

The dictionary defines a road as any paved or cobblestoned path that gets you from A to B, where vehicles, people, or animals can travel, meaning that off-road is all that doesn't fit that bill - therefore, anything that's not paved, cobblestoned, or generally manicured and maintained.

So we need to get something straight: what counts as off-road isn't up for debate, it's not emotional, it's a fact, not an opinion. Whether it's the path Uncle John takes to get to his hillside farm or the giant dunes a Dakar racer blasts across, it's all off-roading, and that's a key point to remember.

It's crucial to remember because not every off-road enthusiast wants to throw down at the Dakar Rally or a BAJA 1000, and that's great news we should all get behind, as a sport can't survive on its elite edge alone. So when a hardcore enduro fan tells someone cruising on an ADV bike that they're not really off-roading, that comment is not just showing a serious lack of empathy and understanding of our sport as a whole, it is also clearly implying that what they're doing isn't legit, or important.

The same could be said about those who diss bikes for being 'overkill' for certain terrains. Again, the empathy and respect shown in those comments are seriously lacking, not to mention the signs of a narrow vision and a lack of broader understanding of our sport. Nobody has their first off-road experience at Erzberg, not all of us started ripping trails on a 50cc at age 4, nor do we all have the same purchase power, at any age range.

In truth, the one thing we share in common as a community is this hard-to-describe love for kicking up dust. So, remembering that there's off-road for every taste, budget, and experience level is key. We're way more likely to encourage someone to try out new terrains and bikes with a little support and understanding, rather than comparing them to what we personally like and can do.

That kind of respect and constructive encouragement is what makes or breaks a thriving, healthy community as opposed to one that's divided and stuck in its ways.

With that mindset, it's important to address some other often-overlooked aspects of our sport and keep them in mind to foster a more inclusive view of this wild, dusty world we all love.

TERRAIN RATINGS

I'll be honest with you - language classes were never my cup of tea, and it took years after my school days to appreciate the real weight of words. A dictionary isn't a hefty book just for kicks; it's chock-full of a vast array of words, each with its own unique meaning.

Even synonyms, while similar and somewhat interchangeable, have nuances that make one fit a scenario better than another.

Yet, in our off-road world, we often find ourselves linguistically strapped when trying to define terrains, defaulting to oversimplified terms like "hard" or "easy."

Those labels are hugely flawed because they don't describe to who a terrain might be hard or easy for, nor do they give any real insight into the terrain's characteristics.

To sidestep the classic blunder of hitting an "easy" trail only to break something because it was not so easy after all, we've got to better craft our own lingo.

Though not widespread, there are some terrain rating systems out there, offering different levels of complexity and applicable to various disciplines, from ADV to hard enduro.

That said, I'm partial to a system that wasn't my brainchild, but that I've taken a shine to. Originally designed for bikes over 650cc on adventure tracks, it can be adapted for the vast majority of bikes in recreational riding, although it becomes a bit more restrictive when you dive into pure enduro or hard enduro. Yet, if there's a community will, one can surely cook up additional ratings for these niches and their unique traits.

The system I speak off, and that I use in my trainings - which you'll find attached - pivots on three main pillars: general considerations laying out some basic rules; the rider's classification across three evolutionary stages; and the terrain rating itself, split into five difficulty levels.

So for those looking to communicate better with their riding friends, offer a more precise language to ride organizers, or even apply this terrain rating method to their own adventures, I recommend a close read and sharing it around.

As a community, we all benefit when we speak the same language and can articulate exactly what we mean by the diverse types of off-road.

TRAIL ETIQUETTE AND SAFETY

Once you peel away from the racetrack, it quickly becomes clear that our off-road world is about much more than just gunning for first place. Our safety, and that of others, along with the unwritten rules of etiquette and common sense that we should uphold, are the bedrock of our sport.

So, let's start with a reminder: it doesn't matter how slick your riding is, how much you've got the technique down, or any other key control point of your bike. If you're not tuned in to what's beyond your "man/machine" bubble, you're not maximizing your safety, nor performance for that matter.

This means paying attention to parameters like weather conditions and forecast, our overall health as a rider, or even the condition of our safety gear. No detail should be overlooked if we intend to day in and day out take the most out of a motorcycle adventures.

With that in mind, it's a no-brainer to say that we should always ride with a helmet and match our remaining safety gear to the type of terrain we're tackling. Like, not wearing street boots for enduro or motocross is just plain common sense and something easy to stick to.

So let's use our time wisely and instead focus our debate on a couple of lesser-discussed topics that don't always get the attention they deserve.

HYDRATION

Hydration is something that for many, only seems critical in competitive settings. Yet, when it comes to leisure riding, many of us live or ride in areas of extreme temperatures, and are no strangers to a cold one during or at the end of our rides, leading us to inadvertently create the proper setting for dehydration, warranting a deeper discussion on the subject.

Considering that most casual adventure and leisure riders tend to not be in top physical shape, might not have a solid grasp of their bike and riding techniques, and yet - a good chunk of them - are on bikes with power-to-weight ratios that are a bit too spicy for beginners and even intermediates, I'd say any leg up you can get is worth taking.

Luckily, and on that note, many health and performance professionals believe that hydration is probably the simplest way to boost our performance in various ways, from general health to ensuring our body is working at its best when we need it to, no matter the level we are trying to push it to. So yes, hydration is, without a doubt, something to keep on your radar, even when you're just out there for the fun of it.

Cramps, dizziness, blurred vision, or a lapse in focus become red flags of dehydration - a non-negotiable reminder that, just as a bike won't run without fuel or oil, neither can we.

So, we can to a certain extent say that hydration is an art in itself, meaning that while we don't need to get into the nitty-gritty that can stay reserved for the competitive scene, a basic understanding will be crucial to positively impact our safety.

First off, we need to understand that staying hydrated is more than just hitting a general predefined water quantity on a daily basis., with that being an oversimplification that doesn't quite cover it. Sure, water is vital for keeping all your systems running smoothly, but true hydration is a balancing act. It's about managing a cocktail of minerals - sodium, potassium, calcium, and magnesium - which you won't get from tap – or spring - water alone.

These minerals are the unsung heroes in our bodies, maintaining fluid balance and supporting muscle and nerve function - making them non-negotiable for any rider, as their good balance keeps dehydration at bay, steering us clear of muscle cramps, fatigue, and a dip in our mental game. With that in mind, balancing these and other minerals through a well-rounded diet is key to proper hydration and overall health.

There are ways to calculate your exact daily fluid needs, accounting for weight fluctuations, sweat rates, body temperature, and more. But that's diving deeper than we need for our current goal.

What's crucial for us, the everyday riders, is to understand that there's a personalized amount each of us needs to drink. While the ballpark figure of about half a gallon a day floats around, that's not a one-size-fits-all solution.

After all, a person who's 5'7" and 110 pounds sitting in an office all day will have different hydration needs from someone who's 6'3", 200 pounds, in full gear, tearing up the trails on a 104°F day.

Keep these key points in mind, and you'll not only boost your riding but also your overall health.

So, it's essential to grasp that on our ride days, we're going to need more water and minerals than on a regular workday. And since hydration in this context is about sustaining performance, we shouldn't neglect it even on non-riding days. That means that drinking two quarts of electrolyte water on Sunday morning before riding out won't do much good.

Hydration is an ongoing process, one that needs to be tailored daily to the weather conditions, our exertion levels, or whatever situation we're in - a task that anyone can manage with a bit of common sense and understanding.

That common sense should also prompt us to limit or control our intake of sugary drinks, alcohol, or caffeine, whether we're out on the trails or off the bike, especially on multi-day adventures. These beverages are diuretic; meaning, they don't help us retain the electrolytes and water we need, on the contrary, they cause us to lose them even faster.

However, it doesn't mean we should just guzzle more water on those days - when it comes to hydration, more doesn't always equal better. While we've learned that water alone doesn't provide all the minerals we need and reaching for electrolyte supplements can be crucial, too much water can also lead to issues, potentially causing a state of overhydration.

This condition dilutes the minerals we actually have in our body, which is a condition that can be deadly in the extreme and is definitely harder to detect and manage than dehydration, given the lack of noticeable symptoms.

With this in mind, maintaining a diet rich in vegetables and fruits - naturally packed with the minerals we need - and knowing how to adjust our electrolyte intake for different demands will keep us out of trouble with ease. Trouble in this case can range from dry mouth or headaches to fatigue, drowsiness, and dizziness - dangerous when riding and also the first signs of dehydration, which in its extreme can also be life-threatening.

Make no mistake: dehydration and overhydration aren't just concerns for competitive athletes - they're universal human issues, whether you're on or off the bike, making it a subject we all need to be mindful of, and one that with just a little attention, can help us maximize our physical and mental performance, ensuring our body always operates at its best.

DON'T GET RUN OVER

Taking a spill is as much a part of off-roading as filling up on gas or changing tires, and faced with this reality, we tend to adopt one of two attitudes when our bike is no longer upright. On one hand, there are those who use a fall as an opportunity to snap as many photos as possible, on the other hand, there are those who, as quickly as possible, do whatever it takes to pick the bike up off the ground before anyone notices what happened.

In a competitive environment, the speed with which we can rejoin the race is crucial, but in a recreational setting, the primary focus is always our safety and, to maximize that, we should use time as a tool.

We must then begin by understanding that a fall automatically triggers a strong hormonal and physical reaction in our body, known to many as the fight-or-flight response. Thus, by using the adrenaline that has been pumped into our blood as fuel, we can easily lift a 440-pound bike off the ground without proper technique, risking serious injuries to muscles and tendons, and even broken bones.

Therefore, the option to stop and take some pictures appears to have clear advantages, as it will give us the necessary time for our body to begin to relax and stabilize, so we can consciously understand if we have any injury masked by the adrenaline spike.

Nevertheless, this does not mean that the first thing we should do is update Facebook or Instagram.

Falls can happen anywhere on the track, which means they can, and often do, occur on blind turns, climbs without visibility, or any other place where oncoming traffic can't see us. With that in mind, it's important to follow a small checklist to ensure our safety while picking up the bike.

First of all, we should turn off the vehicle and make sure we are intact, which is easy to check by running our hands over our body and taking some deep breaths to lower our heart rate. In the case of more aggressive falls - but not immediately incapacitating ones - removing the helmet, gloves, and jacket is advisable.

The time spent in this process will not only help us rebalance hormonal levels, blood pressure, and heart rate but also unveil any potential injuries that could be exacerbated by attempting to lift the bike.

Ensuring we are intact, preferably off to the side of the track, it's then time to secure the scene, especially if you find yourself in an area with limited visibility.

Since motorcycles aren't equipped with emergency triangles or other kinds of hazard signs like cars, if we're riding solo, we can position our helmet and jacket at either end of the crash site to alert oncoming traffic that something's amiss. It's not foolproof, but it generally causes riders to at least slow down.

If we're not alone, we can take further measures by using the bikes at our disposal to block off the section where the incident occurred. Placing a motorcycle crosswise with its hazard lights flashing at each end of the section acts as a clear signal to others that there's an issue ahead and that they should at the bare minimum decelerate, if not come to an immediate stop.

At this point, having used our time to ensure our own physical integrity and the safety of the scene - measures that will greatly reduce the risk of being run over during the next step - all that's left is to right the motorcycle. If we have friends with us, this should be done with their assistance.

Riding bikes, regardless of the discipline we choose to embrace is an individual sport usually carried out in a group, in the same manner it tends to be a marathon, not a sprint, so asking for help replaces the significant individual physical strain of lifting a bike alone with a more manageable and dispersed collective effort.

FIRST AID KIT

The potential for injury while riding is undeniable, so we face a choice: hope for someone to come to our aid when an accident occurs, or be prepared to take action ourselves, whether for us or for someone else, rider or not.

With this in mind, I always recommend that all of us as riders take a first aid course, and if possible, one tailored specifically for motorcycle incidents and accidents in remote locations, aka off-road. This training should equip us with the skills not only to resuscitate someone, but also to manage simple cuts, for example. And I say specific courses for riders because these tend to be more comprehensive than "mere" first aid courses, covering issues like detecting internal bleeding, helmet removal, stabilizing broken bones, and some even covering topics like dealing with field amputations.

Bearing that knowledge, regardless of the course, what's important is not only acquiring the basic knowledge needed but also learning to build and maintain our own medical kit, an essential piece of gear for any adventurer to keep on their bike at all times.

I say build our own kit because, regrettably, there are too many riders equipped with pre-purchased kits who either don't know what's inside, because they never opened it, or don't know how to use everything in it, making its existence a matter of faith that someone nearby will.

This is quite similar to what happens with field mechanical kits, for example. However, medical or first aid kits have an additional step not required for mechanical kits: they include components with expiration dates.

Moreover, just like field mechanical kits, first aid kits can and should be tailored to our specific needs. For instance, while Paracetamol is a staple in my kit for any ride or international trip, Imodium is reserved for countries where local cuisine or water might introduce intestinal issues less common back home.

Similarly, the emergency equipment necessary for an area where an ambulance or rescue helicopter can arrive within 20 to 30 minutes is not the same as what's needed in a location where help may be days away. Thus, adapting to our real needs ensures we're not overloaded yet still prepared to handle the most typical problems of our activity across any terrain, location, and climate.

Moreover, the mere fact of having a kit we know how to use can make a difference not just to our safety but to anyone we might encounter on the track, on a bike, or off it.

Yet, there are a few more steps we can easily undertake that will greatly assist in times of need. The first is, unquestionably, maintaining proficiency through refresher courses, as the main goal is never having to use what we've learned, which leads to a loss of the procedural memory needed to act effectively when necessary.

With that said, to ensure I always have a method to help me get into the right mindset and prevent major mistakes, besides keeping my medical kit visible or informing my companions of its location, I ensure that inside there's a cheat sheet, immediately visible upon opening the kit. This cheat sheet - in Annex II - allows me to quickly perform the necessary triage while unlocking the memories of my training, also enabling anyone who might need to use the kit on my behalf to at least relay accurate information to emergency services.

Safety is a mix of training and preparation, and our medical kit is our last line of defense in the field, so it should be taken as seriously as any other aspect of our off-road universe.

SHOW THE RESPECT YOU WANT TO RECEIVE

Now that we've covered how to safeguard our physical well-being, it's vital to remember that a public trail (or lack thereof) is not a racetrack, even in the most remote corners of the globe. This means that trails are often used by other people, whether they're on foot or operating leisure and professional vehicles, like tractors, while also being home to wildlife of all shapes, sizes, and speeds.

That's why, despite the lack of signage on most trails, they are not only still governed by traffic rules but also by common sense. Simple actions like slowing down when passing people or houses, obeying no-entry signs or "please close the gate" requests, not trampling over crops, and exchanging a word or two with people we meet on the trails make a world of difference in the court of public opinion, as well as in the enjoyment of our sport.

This enjoyment is increasingly being restricted in Europe and beyond, with some countries confining off-road riding to specific private tracks and predefined routes. Generally speaking, to ensure the continuation of our sport, we just need to show others the same respect we'd like to receive, even if it might disrupt our riding rhythm.

Once we've ensured our safety and deployed the necessary and deserved respect for the trails and communities that live along them, the only thing left is to ensure we respect the nature where these trails are located.

Few experiences are as rewarding for adventure riders as setting off into the sunset with a tent on the bike with the intent to not return home for a day or two.

While there's nothing wrong with this concept - quite the opposite - when we move from the planning stage to reality, we must take certain precautions if we want to continue being able to have such adventures. In many parts of the world, this freedom is highly encouraged, but in others, like Portugal, where I'm from, it is heavily regulated. Therefore, ensuring the legality of our chosen camping or picnic spot is crucial.

This has implications such as making a fire for a meal might be illegal in many places and at certain times of the year, and while we can be an excellent resource for authorities as early fire warning scouts, we must avoid accidentally starting one due to not knowing local rules and guidelines.

To conclude, and on the highest note of respect for the nature we love to ride in, we should always try to leave all places better than we found them. So, ensure you have a bag with you, not just to make sure you don't leave any trash behind but also to pick up any litter you might encounter along the way to dispose of properly later.

By respecting ourselves, others, and following simple common-sense gestures, we not only ensure a healthy future for our sport but also, and perhaps more importantly, that it has a future to start with.

CHAPTER II

THE RIGHT BIKE

Remember your first driving lesson in a car? If you did it in a school or with a more safety biased parent, they stressed adjusting the seat and mirrors to fit your body before even turning the key - or at least, that's what was asked of me when I got my license. For motorcycles though, the adjustment advice often stops at the mirrors - if that much - a simplification that doesn't do justice to what we as riders need to actually consider for a proper riding position on our bikes.

Over the years, I've found myself drifting away from certain bikes because they just didn't fit my frame, having gravitated towards models that naturally offer me more comfort, control, and safety. But this doesn’t mean you should do the same if your current bike doesn't feel entirely comfortable. If you're happy with your ride, even if it's not ergonomically perfect, that's fine, as long as you're aware of what this means for your own safety.

My advice isn't then about limiting who can or should ride certain bikes. It's more about understanding whether your bike-rider combination is ergonomically compatible, keeping in mind that a bike's design, power, or specs on paper don't necessarily make it the safest or most effective choice for anyone.

This issue stems from the fact that motorcycles, designed for a broad range of riders, need to be fine-tuned to our unique and specific dimensions if we wish to use them correctly.

Consider this: the same bike model can be bought by someone who's 5'3" or 6'7". Everything from the handlebar size, which might need to be adjusted for arm length, to the type of seat, footpeg position, handlebar height in relation to the triple clamps (where the suspension forks attach to the bike), and even the bike's ground clearance, are all factors that should be considered for a comfortable and safe riding position - just like in a car.

With such an extensive checklist, it's clear that ergonomic issues in motorcycles come in more flavors than just bikes being too tall for a specific rider, as problems faced by taller riders are just as real and detrimental to good riding.

Thus, it would be incorrect to assume that there's a clear, direct solution for all ergonomic adjustment issues. Since these solutions are a personal compromise and, while they may follow certain patterns, adjustments become more of an art than a science.

With all this in mind, this chapter is dedicated to highlighting what we should consider, rather than offering direct solutions.

FROM THE SEAT TO THE GROUND

Most riders under 5'11" have likely encountered an off-road motorcycle - be it an adventure, enduro, or dual-sport - where it was nearly impossible to comfortably plant both feet on the ground.

This reality, while factual, often misses four crucial nuances, which can leave us puzzled when we find it easier to touch the ground on a bike with a taller seat than on a shorter one, or when we wonder why a friend of similar height doesn't face the same tip-toe issues on bikes that to us are a challenge.

Firstly, even though human proportions between the torso and limbs are relatively consistent, not everyone of the same height has identical leg or arm lengths. Take Olympic swimmer Michael Phelps, for example. Standing at 6'3", he shares a leg inseam of around 31.9 inches with Hicham El Guerrouj, an Olympic runner who is just 5'5" tall. So, when considering seat height relative to the ground, it's crucial to measure the distance from our inseam to the ground, not our total height, because it's our leg length that counts here.

Secondly, there's a common misconception that both feet need to be flat on the ground to ensure safety on a bike. Geometry dictates that only three points are needed for a stable plane, and a plane, by definition, is a solid, stable structure.

As such, with the bike's two wheels on the ground - if all is according to plan - we only need one foot firmly planted to ensure the stability of the bike-rider combo when stationary. Don't get me wrong, it's undoubtedly more comfortable to have both feet grounded - after all, we're bipeds by nature - but natural doesn't mean impossible or even uncomfortable.

Thirdly, we must consider the width of the seat, a detail rarely mentioned in any motorcycle specs. If you measure 27.6 inches from inseam to the ground with your feet together, basic geometry tells us that spreading your feet apart will reduce your distance to the ground. This explains why many riders find it easier to sit on an enduro bike than on an adventure bike, even though both might have the same seat height.

Lastly, we often give up on a bike based on a decontextualized reading of the manufacturer's measurements. If the width of the seat and the specific leg measurements directly impact our ability to reach the ground, and if choosing to use one or both feet changes our possibilities, then the sag of each bike is the final ingredient in this equation.

Sag - a critical measurement that should always be considered and should always be present - is the change in seat height between when we sit on the bike as opposed to its unladen position. This simple change can turn a bike that seems prohibitively tall on paper into one that's just perfect.

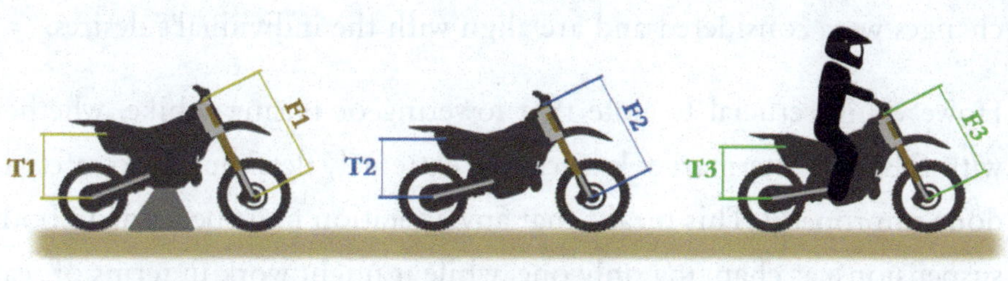

Image 2.1 – The various sag measurements. 1 - bike sitting on a stand with the wheels unweighted. 2 - bike resting its own height on the wheels. 3 – bike's weight plus rider with gear resting on the wheels.

This highlights why it's crucial to always try out a bike before making judgments about our ability to handle it.

That being said, if all else fails and your desire to own a specific bike remains steadfast, there are still options.

Lowering kits - or ones for raising the bike - are incredibly effective, with most manufacturers offering factory options within acceptable maximum and minimum heights before the bike's characteristics are completely altered.

However, even if these modifications drastically change the bike's characteristics, that doesn't mean they're directly harmful, meaning only that they can become incompatible with the intended purpose of the rider. Raising or lowering the bike alters its center of gravity, wheelbase, and available suspension travel, among other things, which indicates that in some cases, these changes may align directly with what you're looking for, making them an asset.

Again, it's vital to remember that each person is unique, not just in measurements but also in needs and objectives. This means that as a community, we should move away from saying bikes shouldn't be lowered or raised and instead discuss whether all implications of such changes were considered and are align with the individual's desires.

However, it's crucial to note that lowering or raising a bike, whether with OEM or aftermarket kits, can be extremely detrimental to safety if done improperly. This means that any alteration must be made to both suspensions, as changing only one, while it might work in terms of seat height when stationary, will drastically alter the bike's geometry and safety while in motion.

FROM THE SEAT TO THE PEGS

To avoid modifying a bike's original dimensions, many riders, from casual to professional, opt to change the seat, choosing a higher or lower one as a method to adjust their sitting height.

While this option is perfectly acceptable and common, its implications extend beyond just altering the maximum seat height, particularly concerning the distance from the seat to the footpegs, as having footpegs too close or too far from the seat doesn't promote good ergonomics, quickly making any ride uncomfortable and affecting the ease of transitioning from sitting to standing.

Knowing this, we could say that changing the seat is a solution to be avoided, but that would eliminate a significant tool for ergonomic adjustment.

What's important to keep in mind is that changing the seat height - or for some, even with the original seat due to their unique dimensions - may necessitate adjusting the footpeg position, whether in height, vertically in relation to the bike, or longitudinally, moving them forward or backward.

That said, moving the footpegs from their original position has significant implications on weight distribution when standing, and even while sitting, meaning that this option should only be chosen after careful and thoughtful analysis and should not be done simply for the sake of buying a new accessory.

FROM PEGS TO FOOT LEVERS

When considering the longitudinal position of foot pegs, for many, a crucial factor becomes the size of the rider's feet.

While the distance from the peg to the brake or gear shifter is fine for most, riders with shoe sizes over 10 or 11 can start running into issues. Although applying good foot positioning techniques helps a lot, sometimes it's just not possible to find a comfortable position on certain bikes due to the lack of available space.

In these cases, while adjusting the pegs is an option, we must remember that it's one that comes with its own specific implications. Like that, our first choice should be longer aftermarket shifters or modifying the existing ones, however, they might not always suffice.

If so, changing the pegs' position might become a necessity, but one that should be done cautiously, as previously mentioned.

FROM THE HANDLEBAR TO THE LEVERS

Just as with the foot controls, the way we reach the clutch and front brake levers is crucial. We don't all have the same hand and finger sizes, so proper lever positioning can make a significant difference in both riding comfort and safety. This means that there are several key points to keep in mind, starting with the lever's position relative to where your fingers make contact.

When off-roading, you're typically only using one or two fingers on the levers, so it's essential to maximize how effectively they operate. Remember, one reason that motorcycle-riding police forces tend to be trained to use all four fingers on the clutch on the road is for energy management purposes, a technique we simply cannot safely use off-road. As such, using just the tips of your fingers might offer greater sensitivity, but it also makes applying force more challenging, especially with certain lever designs, forcing us to have a deeper look into this issue.

Positioning the lever on the phalanx just below the last bend of your finger decreases sensitivity slightly but allows for more of your finger to wrap around the lever, making it easier to apply force. This balance is key in off-road scenarios, where quick and efficient lever manipulation is often required.

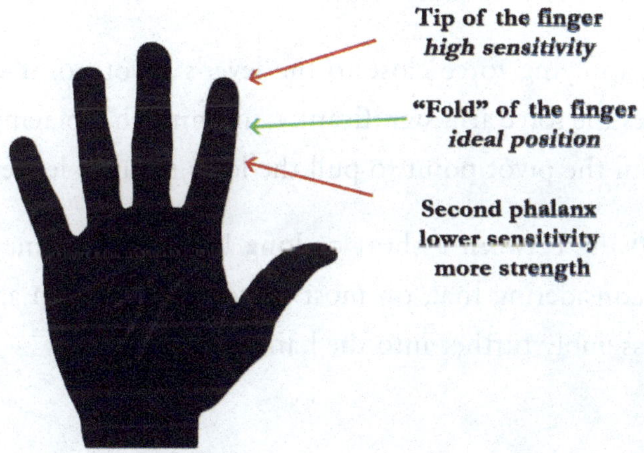

Image 2.2 – Lever positioning on fingers.

Using these concepts, aiming for the area between the distal and middle phalanx - or the last bend before the fingertip - to operate the lever will yield the best results. In some cases, a lever with adjustable distance might be necessary to achieve this optimal position.

However, it's important to note that achieving this finger position with a regular or short lever is less significant; as what we're seeking here is the maximization of our biomechanics through proper finger placement.

For those using cable clutches, good maintenance of the cable and clutch can significantly reduce lever stiffness, and in turn, hand fatigue, but pulling the perch closer to the center of the handlebar offers an even greater and cost-free advantage.

We again need to keep in mind that physics tells us through leverage formulas that the longer the force arm, the less application of force is required to achieve the same result.

Therefore, applying force close to the lever's pivot point versus at its end changes the force arm significantly, meaning that placing our finger farther from the pivot point to pull the lever requires less effort.

Here, choosing between a short or long lever can become important, especially considering that, on most motorcycles, we can always adjust the lever assembly further into the handlebar's center.

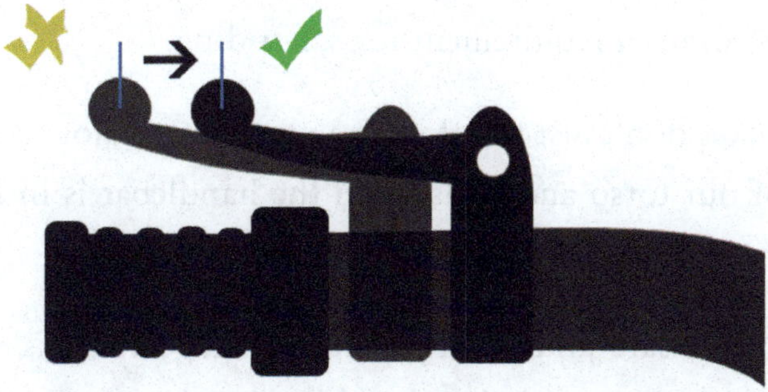

Image 2.3 – Possible improvement of control and sensitivity by changing the perch position on the handlebar.

Even so, as the end of the lever is the furthest point from the handlebar, riders with smaller hands may need an adjustable lever to take advantage of this tip, for both the clutch and brake, that follows the same physical principles.

FROM THE TORSO TO THE HANDLEBAR

While adjusting levers is crucial for optimal bike operation, if we can't reach the handlebar to begin with, we can't operate anything at all.

This might seem like an odd statement, as while we've all seen people struggle to ride a bike because they can't reach the ground, it's rare to hear of someone unable to start riding because they can't properly reach the handlebar. However, as statistically correct as that statement may be, it doesn't mean that everyone we've seen riding out should have done so.

This is because merely reaching the handlebar with our hands is far from the minimum requirement for safe riding.

After all, bikes don't just travel in straight lines, so not considering the position of our torso and arms when the handlebar is in full lock can be a costly oversight.

The goal for a safe and comfortable riding position is reaching the handlebar with our arms relaxed and bent at the elbows while keeping our torso straight an upright. This allows us to extend our arms to move the handlebar from side to side without drastically altering our torso's position, if at all.

It's undoubtedly possible to ride bikes without achieving this position, but that doesn't mean the ride will be comfortable, safe, or allow for any kind of technical consistency, even with a deep understanding of the risk involved.

Using risers or even adjusting the angle of the handlebar are excellent methods to resolve this issue. However, these are not solutions without consequences, making their application a careful job that deserves more respect than it usually gets, especially when we talk about risers.

WEIGHT

After all these adjustments, we need to talk about the bike's weight. This isn't about the potential to alter it within certain parameters, but rather about safety and ergonomics, with the critical question being: Can the rider lift their own bike alone if it falls over, or not?

It might sound like a silly question, after all, who buys an off-road bike they can't lift? Unfortunately, the answer is: a lot more people than we would initially think.

For those who ride adventure bikes mainly on road, this issue might seem somewhat irrelevant. After all, the weight only becomes problematic in certain situations, which can generally be managed with good technique. And if all else fails, there's usually someone around to lend a hand. Off-road, however, where falls are an integral part of the sport, not being able to pick up your bike alone, or even with help, can lead to serious problems.

That's why I always recommend choosing a bike you can lift by yourself at least two to three times in a day and learning the best techniques to do so, both solo and with assistance.

Why two to three times, and preferably consecutively? Because someone who's riding off-road that struggles to lift their bike is generally less technically skilled. So, assuming they're more likely to fall than an experienced rider, they will need to be able to get out of trouble alone, more than once a day. Plus, even with help, poor technique can quickly lead to injuries, which not only hinders your ability to lift the bike in the first place but also increases the likelihood of another fall.

Therefore, considering the weight of the bike, along with the weight of the gear you plan to carry regularly as well as the type of terrains you want to spend most of your time on, is undoubtedly crucial for any rider's safety.

POWER

It goes without saying that power doesn't directly relate to ergonomics. However, since it's something almost all riders tend to tweak, I feel somewhat compelled to mention it, because, in the end, power has direct implications on our safety, especially off-road.

To understand this, let's take a deep breath and critically analyze the facts, setting aside personal feelings.

Today, most new bikes come with electronic aids, with the majority of their off-road modes, reducing engine power in one way or another, and that, is a fact. It's also a fact that many top professional athletes in recent years have chosen not to alter their engines' power for competition all that much, with some even requesting less power rather than more when we start talking about big adventure bikes.

Keeping that notion at hand one must ask: if the world's best riders believe that most bikes are powerful enough to be competitive using the same engine any of us can buy off the lot, and if the pricey electronic aids we get partly serve to reduce horsepower or the way it's delivered to the rear wheel in their off-road settings, then do we, as leisure riders, really need more power or a quicker delivery of it?

Aside from the need to boost one's ego with irrelevant numbers or a kind of pseudo-heroism tied to a questionable lack of self-love, the answer is a simple and resounding no.

There's a clear difference between useful power and maximum power, and brands, through their marketing departments, have been bending over backward to convince us that "more" is unequivocally better, often choosing to overlook the importance and necessity of riding within our own limits.

I always recall with a smile a training session I had with a two-time world supermoto champion, as he explained how, on his BMW GS, he chose to ride off-road with the engine map in rain mode, as this allowed for smoother power application and made it easier to lessen the danger of a mistake or distraction. This approach, adopted by someone with more than enough track-record of his power and speed control, should be applied by all of us, especially when we're just starting out.

Even at more advanced stages, power adjustments made by experienced and professional riders aren't focused solely on huge increases in numbers. Rather, in many cases, they're about tweaking where and when that power kicks in, again emphasizing the concept of useful power. Sticking within that useful power is what will allow us to develop our skills mile after mile and ensure that a momentary distraction doesn't end up with us riding way over our heads.

So, it's crucial to keep in mind that before splurging on power maps, exhaust systems, airbox mods, or even quick throttles, we should first learn to use all the power we already have.

Remember, the goal of our sport is to make it through the day in one piece so we can safely return to our families, and be able to hit the trails again the next day. It's not about being the first to finish the track or the fastest in a ride or adventure where timekeeping only matters for knowing when our vacations or weekends are over.

CHAPTER III

TRACTION AND GEOMETRY

We've all heard about traction, and most likely, even used that term to describe one situation or another. But do we really know what we're talking about when we mention traction? And if we're not entirely in the know, is there value in changing that?

As far as I'm concerned context is key in life, and this topic is no exception. As I mentioned at the outset, the goal of this book is to provide you with as much information as possible in a clear and replicable way. This means that while I strive to be as precise and concrete as possible, sometimes I need to simplify certain concepts or explanations to convey an idea or clarification more effectively. I do this because, at the end of the day, this book isn't a lecture or an academic dissertation; it's a way for you to better understand our two-wheeled off-road world or, at the very least, to introduce concepts that might be new to many, so you can choose to learn more about them if you wish. Chapters I and II are prime examples of this approach.

With that in mind, this brief introduction serves as an apology to those who crave scientific detail, as in this topic about traction, I'm going to have to do just that: simplify concepts and, somewhat nonchalantly, present certain ideas as absolute truths.

Although that may seem wrong, I must stress that the intent here is, like in the rest of the book, to use broader knowledge to help you eliminate fears based on preconceived notions. This will enable you to understand limits that will change the way you approach personal challenges, technical maneuvers, and terrain peculiarities from a more advantageous, positive, and productive standpoint, leading to safer and more consistent riding.

KAMM CIRCLES

Tire development is a billion-dollar industry, and among the reasons for this is the complexity of a tire's functionality. Temperature, pressure, the properties of the rubber used, the nature of the terrain the tire rolls on, deformation under use, bike weight, rider weight, tire dimensions - the list of variables is immense. Obviously, with such an extensive list of details, any minor change significantly impacts the final product and our use of it.

This is why we can leave the deep understanding of each of these variables to those who develop tires, squeezing for ourselves just the required bit of the juice from this orange. In practical terms, this means understanding some concepts that will change how we view tires and understand traction - something much more relevant to our everyday riding.

The first concept to grasp is how much we can or cannot push a tire before it loses traction, a common fear among riders of varying experience levels, especially when it comes to the front tire. The simplest way to understand this is by using a Kamm Circle, traction circle, or any of the other names it's sometimes given.

To comprehend this concept, imagine a circle divided into four parts, with each quadrant representing the limits of acceleration, braking, turning left, and turning right. Thus, we can easily draw boundaries and visually understand tire behavior.

For this example, we will be using a right-hand turn under acceleration.

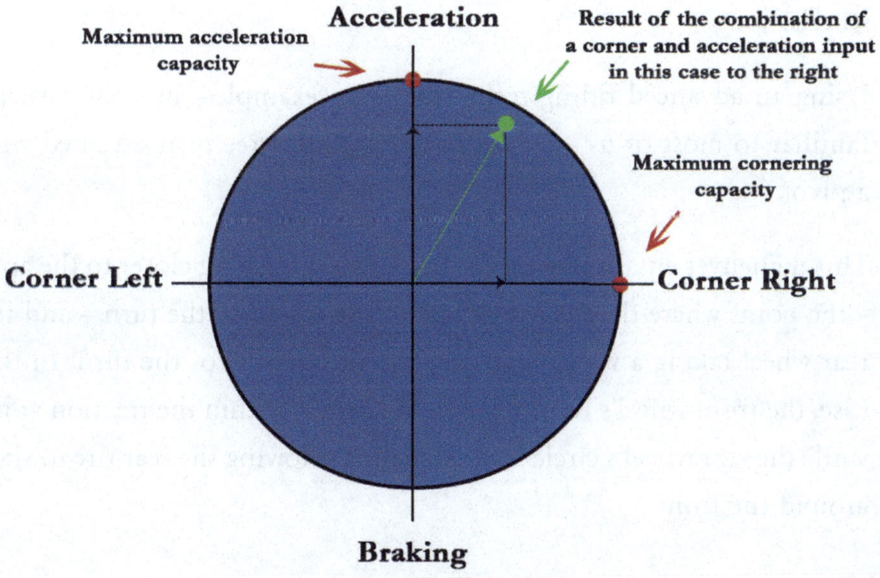

Image 3.1 – 1st Kamm Circle.

When analyzing the graph, we understand that, in this situation, if we want to accelerate more, we'll have to reduce the angle of our turn if we aim to stay within the circle, as stepping outside of it would mean leaving the traction zone and beginning to skid.

On a motorcycle, this concept might be trickier to visualize, but if we think about our four-wheel experiences, we can probably recall at least one turn where we either entered too fast or turned the steering too sharply and drifted off our desired path.

Using the Kamm Circle as a visual aid and our memory as a sensory guide, we can easily understand that by overdoing one of the variables, we exceeded the tire's capabilities, making traction loss inevitable, and as such, missing our trajectory.

Now, let's simulate a practical application of this explanation on two wheels.

Using an advanced riding technique as an example - but one visually familiar to most of us - let's consider a 180-degree turn executed with a pivot.

This maneuver entails having the front wheel tracking closer to the apex - the point where the bike is closest to the inside of the turn - and the rear wheel taking a wider line closer to the outside of the turn. In this case, the front wheel's Kamm Circle is entirely within the traction zone, while the rear wheel's circle is outside of it, allowing the rear tire to slide around the front.

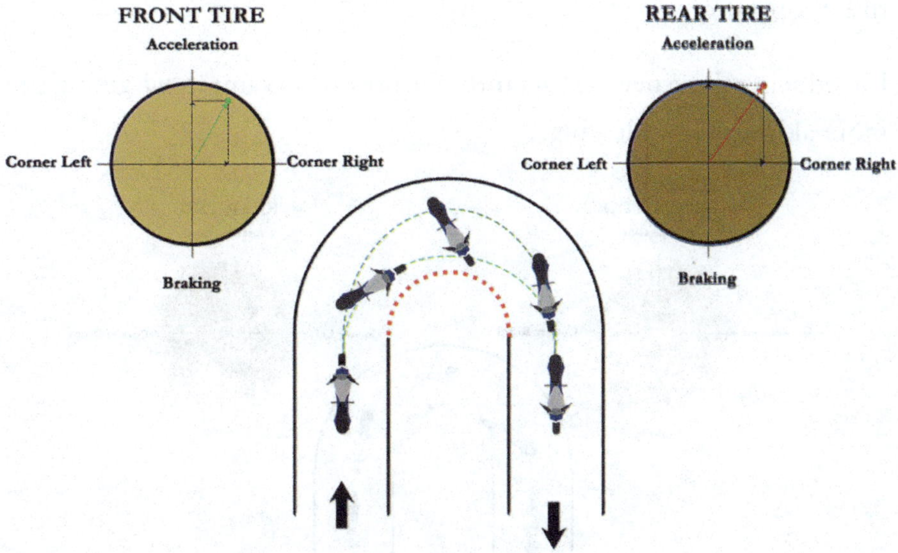

Image 3.2 – 2nd Kamm Circle.

However, in reality, things are a bit more complex than they first appear.

This is because the limits of a Kamm Circle aren't a fixed line but more like a gradient, as traction loss isn't instant but occurs in a relatively progressive manner for the most part. This gradual loss is what allows top riders to operate at the limits of traction. With a bit of training, some sensitivity, and considering how tires are constructed nowadays, it's possible to feel when a tire starts to slide, even to a newcomer to dirt riding.

Another key point to understand is that the a traction circle isn't actually a perfect circle but more of an oval shape. This is due to tires generally having more longitudinal load capacity than lateral, a characteristic that arises partly due to the physical deformation of the tire in operation.

Introducing these new factors into our previous example, we get a graph more akin to the following.

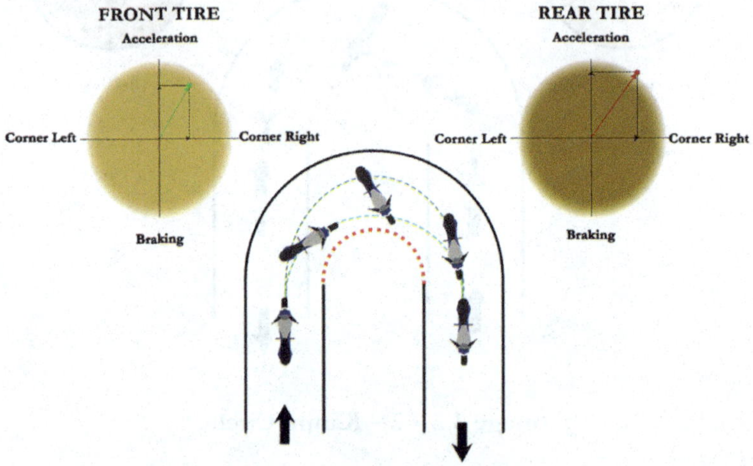

Image 3.3 – 3rd Kamm Circle.

Even though there's a saying that it's not the size that matters, but what you do with it, in this case, both the size and what you do with it matter. This is because the friction of a rubber tire against the ground is a unique combination of adhesion, deformation, and wear.

Adhesion is a relatively simple concept - think of glue and remember that a cold tire has less adhesion than a hot one, for example.

Regarding deformation, reducing tire pressure is a well-known technique among off-road enthusiasts because it allows the tire to deform more, increasing the contact area with the ground. In practice, as we have a rubber with adhesive properties now contacting a larger ground area, we then have more traction.

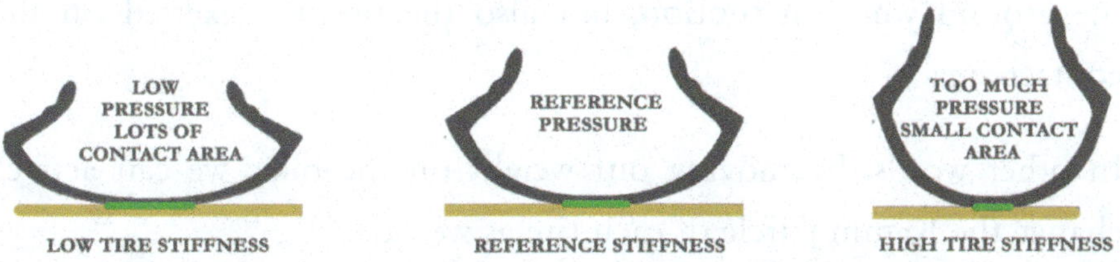

Image 3.4 – Changing tire pressure alters the contact area with the ground.

However, it's important to note that while decreasing pressure increases traction through a larger contact area, it also impacts other factors like shock absorption capacity, sidewall strength, rolling ability on certain terrains and at certain speeds, fuel consumption, and more.

That being said, and since adjusting pressure on the fly to increase or decrease the contact area isn't feasible, what we can do to leverage this principle without affecting other unwanted aspects is to shift our body weight.

Motorcycles are precision-engineered vehicles, so the way we move on them is not only crucial but should equally be precisely calibrated, as with minimal movement, we can significantly impact any bike's behavior.

To understand this concept, we need to consider that about 60% of our body weight is located above our hips. Therefore, someone weighing 220 pounds fully geared, by moving their torso just a few inches forward, backward, or sideways, is effectively shifting 132 pounds of weight of axis.

This means that by shifting more or less weight onto each wheel or footpeg, we're not just actively altering the contact area of our tires with the ground while in motion, but also the pressure exerted on that contact area.

In other words, by moving our weight on the bike, we can actively change the Kamm Circle of each tire as we ride.

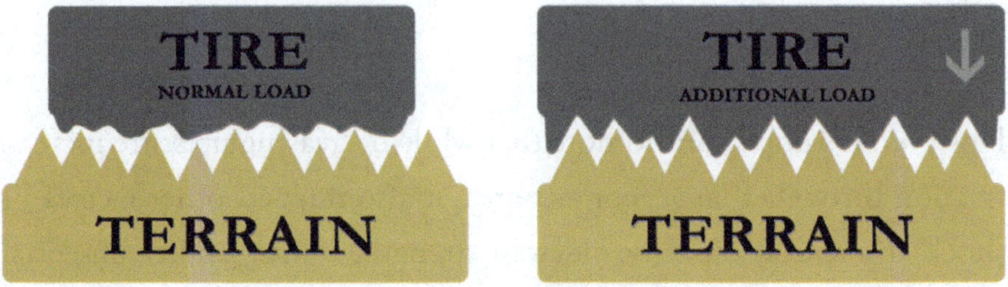

Image 3.5 – Shifting weight onto a tire changes its contact area and deformation on the ground.

Now that we have a bit more context, let's circle back to the pivot turn example.

Generally, the best body position for this maneuver involves leaning forward, shifting more weight onto the front tire and less on the rear.

At this point, it shouldn't be hard to understand that this body shift takes advantage of the tire's capacity to deform under load, effectively altering each tire's contact area with the ground. However, it's crucial to grasp that there's a physical limit to the correlation between applied weight and traction gain - meaning, piling a ton of force onto a tire won't result in a ton of traction, a critical point always important to keep in mind.

Without getting too bogged down in the technicalities, if we revisit the Kamm Circles of our pivot turn, we now have a fuller understanding of what's happening to our tires during the maneuver, considering our body weight shift.

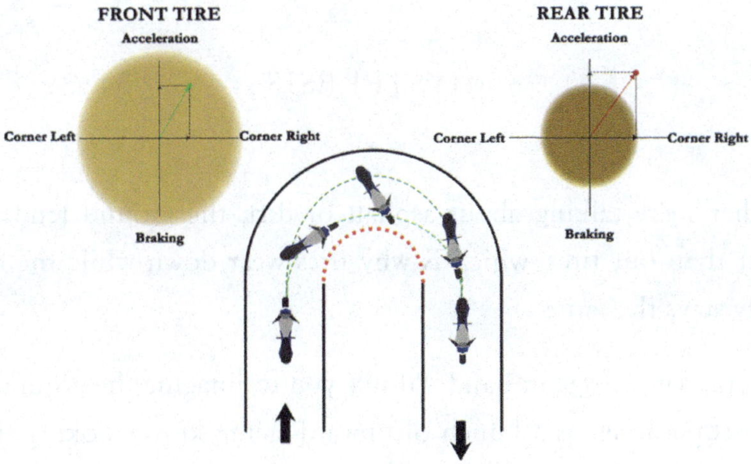

Image 3.6 – 4th Kamm Circle.

I chose to discuss a pivot turn, but everything we've talked about here directly applies to other maneuvers, which becomes clear when we consider the different body positions required in various situations.

For example, during emergency braking, we tend to shift our weight backward to counter the forward energy transfer and avoid overloading the front tire, especially under emergency braking.

The same logic applies to descending steep hills. Conversely, on ascents, we might need to move forward or backward to adjust the traction on the rear wheel.

These and other body movements exist out of pure necessity to alter the Kamm Circles while in motion. Understanding this begins to explain why off-road riding, with its primary focus on maximizing traction, is such a physically demanding and highly dynamic sport that is not easy to master via trial and error alone.

HYSTERESIS

Whether we're talking about asphalt or dirt, the ground tends to be harder than our tires, which is why tires wear down while the terrain mostly stays the same.

With this knowledge in hand, I'd like you to imagine the ground, on a microscopic level, as a bunch of upward-facing knives poking the tire every time they come into contact.

Also, keep in mind that tires are viscoelastic, meaning they tend to return to their original shape. This property is evident when you press your nail against a tire and observe how the indentation disappears as the tire resumes its original form. The difference in energy expended during the compression and recovery process of the rubber is known as hysteresis.

A tire with higher hysteresis will conform more precisely to the ground, gripping the 'knives' of the terrain better and enhancing traction, but this also leads to faster overall tire wear.

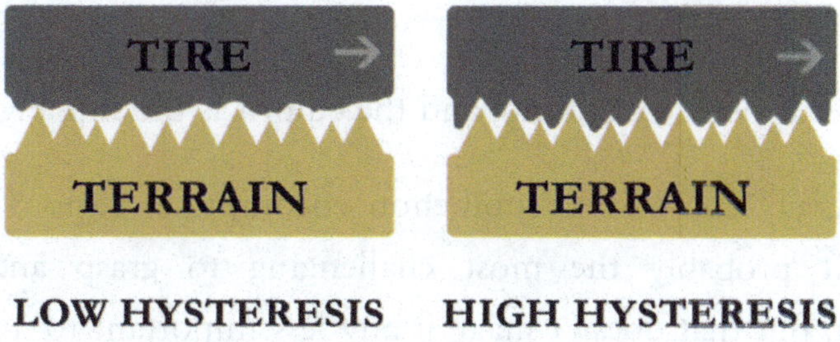

Image 3.7 – Demonstration of hysteresis.

With this visual image of knives and a basic understanding of hysteresis, it becomes relatively simple to grasp why a softer tire tends to offer more traction but also wears out quicker than a harder tire.

Circling back to when we said earlier that a ton of weight doesn't translate to a ton of traction, hysteresis is one of the culprits.

This is because when a tire fully compresses against the ground due to weight, the excess energy given by the extra weight still needs to be dissipated. So, rather than resulting in increased traction (which is already maximized by that point), it leads to effects like deformation of the tire's carcass, among other unwanted effects. This allows us to say that at a certain point, if we put too much weight on a wheel, we'll lose traction due to excessive tension.

TIRE DYNAMICS IN A TURN

One image we all have from the world of riding is MotoGP bikes sliding into a turn, seemingly sideways. But are our eyes somewhat deceiving us?

The answer is both yes and no, and the culprit is the slip angle.

This relatively complex phenomenon encompasses various concepts, making it probably the most challenging to grasp and explain succinctly, but that doesn't make it any less important to understand, even at a basic level.

In brief, the slip angle is the angle formed between the direction the tire is pointing and the direction the bike is actually traveling, a simple enough notion to work with.

However, it's crucial to note that, besides the slip angle, we also have a slip area - concepts that are similar and occur simultaneously, yet are distinctly different. The latter refers to the part of the tire that is slipping within the contact patch with the ground, a more abstract concept to grasp, but still, one worth knowing if we wish to have a full picture of what is happening.

The simplest way to understand this slip area is to think of worn-out Velcro that can slip between its hooks and loops with some force, failing to provide the fixed contact expected of the joined Velcro. In the case of tires, this force of union is known as traction.

Adding to this list of concepts, we have the cornering cone, which, for all practical purposes, dictates the angle of lean necessary for a bike to take a certain turn with a specific tire profile and speed.

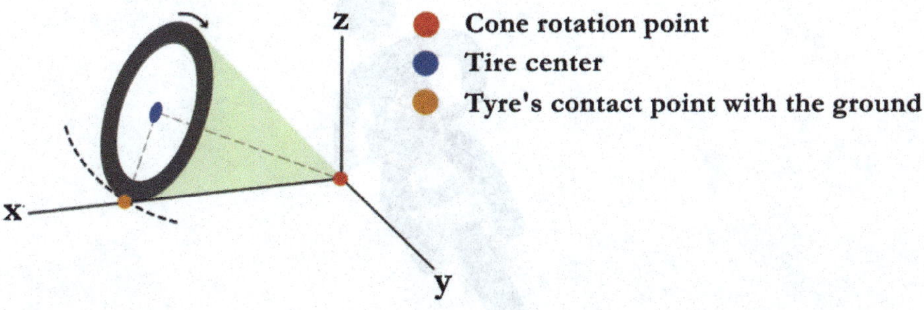

Image 3.8 – Cornering cone.

It's also essential at this stage to talk about countersteering, a maneuver known to some, used by all, but not necessarily actively initiated by every rider.

This is because, unlike a car where turning the wheel steers the car directly, on a motorcycle, we can't turn in the exact same way. In reality, we turn by leaning the bike, and we achieve this lean by briefly steering the handlebar in the opposite direction of the turn we want to take.

The action of the centripetal force at the wheel's contact patch creates a gyroscopic force on our body and bike, causing it to fall in the opposite direction to where the wheel is pointing.

Image 3.9 – Countersteering.

However, when we lean the bike into a turn, we're introducing an angle of inclination on a moving mass, which tends to turn the handlebars in the direction of the curve.

This effect is enhanced by a phenomenon known as gyroscopic precession, along with the bike's trail, which naturally aligns the front wheel. These physical phenomena that align the front wheel lead many to believe that we never actually steered the handlebar outward to initiate the turn in the first place.

This explanation is oversimplified and, in some ways, not entirely accurate when considering the depth to which these topics can be explored. However, our goal here is to understand the basic principles that enable our vehicles to function effectively, nothing more.

With this in mind, despite all the complex physics and engineering, we can create a visual image of these critical concepts to dispel some preconceived fears.

Visually, we also find it easier to grasp and accept the application of certain body positioning and bike control techniques that we will discuss later.

If at this point you are feeling disheartened and confused, don't worry, I promise there is method to the madness at play here.

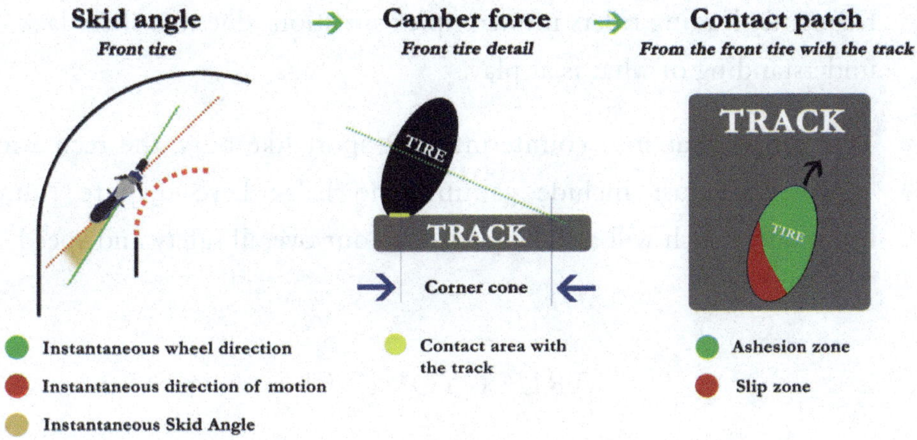

Image 3.10 – Tire dynamics in a turn.

Case and point is if we consider that a common fear to many is thinking that, just because we're on dirt, we'll immediately lose traction, especially in the front. Now we understand that this isn't true, as traction loss is gradual, and not immediate.

We also learned that by mastering the slip angle and Kamm Circles of our tires, we can maximize our traction, allowing us to effectively slide without completely losing traction or control of our bike.

These concepts also help us understand that both on and off the road, we need to lean the bike and use countersteering to allow the motorcycle to function properly, enhancing our safety in the process.

We also realize why the concept of pushing down on the outside footpeg during a turn - whether sitting or standing - reduces the likelihood of the tire slipping.

With all this analysis, we can assert that preconceived notions about certain dangers often stem from contextless tips or explanations, frequently leading riders into complex situations due to a sheer lack of understanding of what is at play.

This proves that in a counterintuitive sport like ours, the recipe for consistency must include gaining knowledge beyond pure riding technique, which will also help increase our overall safety and speed.

TRUST YOURSELF

After all this explanation, many of you might be wondering what to do now with this information, a valid question with a straightforward answer.

By understanding the concepts we've discussed, we more easily grasp what phrases like "shift your weight forward" mean, making their execution more intuitive and controlled.

Another benefit is the realization that a bit of sliding can mean maximizing speed, something not to be confused with a total loss of traction, something that might look spectacular in photos but that can drastically reduce both performance and safety.

To these benefits, I'd add another extremely important one: understanding that there's no such thing as "the best tire" that many seek on social media and among friends.

The best tire depends on a myriad of factors, with the most significant differentiator being ourselves. The type of use we give the tire, the conditions we ride in, our expectations, and even our riding style are factors that drastically affect the performance of any tire we get our hands on. That's why a tire that is brilliant for some is deemed terrible by others.

LET THE FRONT END WORK

Having talked about traction, we must also address geometry, which obliges us to talk, albeit very simplistically, about rake and trail, measurements that help keep our bike upright and which many of us, out of ignorance, consistently prevent from functioning.

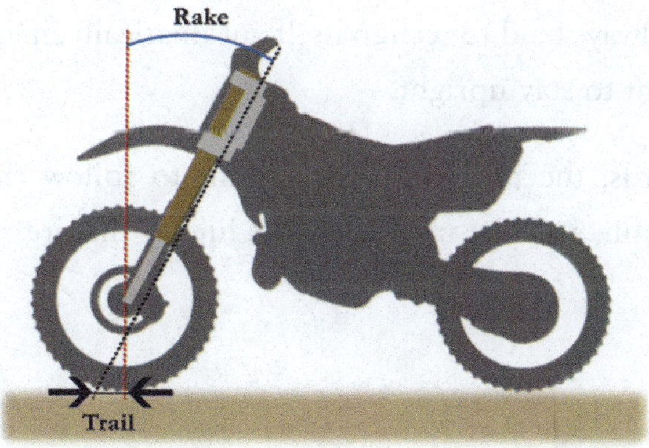

Image 3.11 – Location of Ground Trail (commonly referred to simply as trail).

Diving straight into it, rake is the measurement between the imaginary vertical line that passes through the axle of the front wheel and the line that passes through the steering column, a crucial concept to understand bike geometry. Rake works directly with the trail, which is the measurement between the point where the vertical line from the axle touches the ground and the line created by the continuation of our steering column on the ground.

Trail is crucial to the geometry of any bike for various reasons. For the purposes of this explanation, we can say that less trail means quicker, more nervous steering, while more trail indicates slower steering but a more relaxed journey.

However, what leads me to discuss trail is the insight we gain when looking at this measurement from a different viewpoint, in this case, from above. By seeing what happens from there, we can analyze and understand that whenever we are faced with a situation that takes the front wheel off its direction of motion, because we have a trail point, the tire will always tend to realign itself automatically, meaning the bike itself will want to stay upright.

In other words, the bike will always want to follow the line of the steering column, and not necessarily the line of the tire.

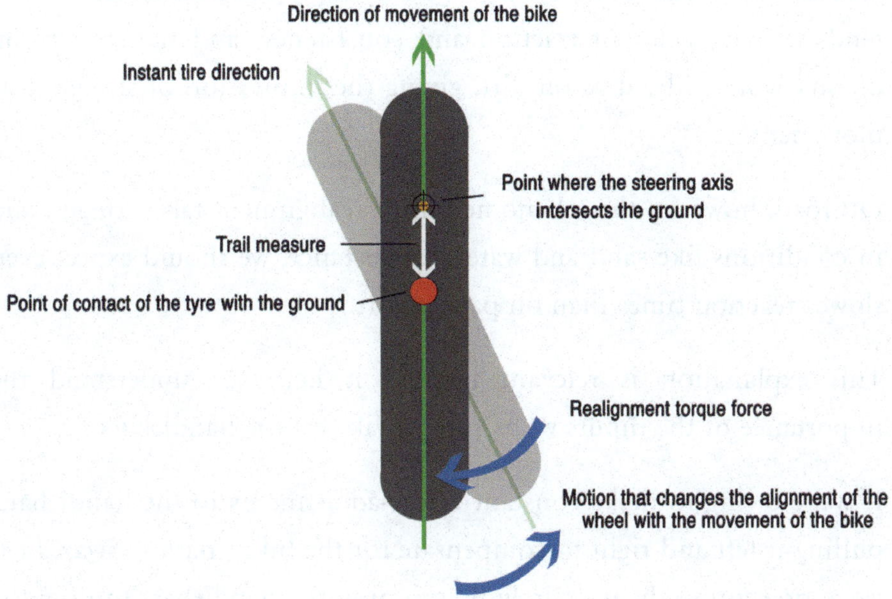

Image 3.12 – How trail works on a front tire viewed from above.

If you've ever watched videos of bikes riding alone because the rider fell off - a phenomenon known to many as ghost riding - one of the major contributors allowing the bike to stay upright is the presence of the trail.

As long as bikes have speed and, consequently, direction, they'll remain upright because the trail, up to a certain point, ensures that the direction automatically stays aligned and, as such, that the bike remains vertical.

We can further expand this concept by stating that, with or without a rider, in reality, no bike travels straight but is instead in constant realignment from left to right, drawing a sinusoidal trajectory. On asphalt, this realignment occurs at a very high rate because the terrain tends to have a lot of traction and consistency, and the suspension doesn't work as hard as on dirt, giving the impression of straight-line movement.

Off-road, however, this phenomenon of realignment takes longer, and in conditions like sand and water, for instance, we should expect even slower reaction times than on packed dirt.

This explanation is relevant because it helps us understand the importance of the inputs we as riders make on the handlebars.

If we're trying to make constant micro-adjustments to the handlebars, pulling it left and right to compensate for the bike's natural sway, or if we're preventing them entirely by gripping the handlebars too tightly, we'll compromise the proper functioning of the bike's geometry and, as such, create blockages in its capabilities of self-alignment and turning.

Thus, by staying calm, waiting, and not gripping the handlebars for dear life, we give the bike time to do what it's designed to do, moving us away from the threshold of a fall, something we'll examine in more detail in the next chapter.

THE WHEEL ROLLS OVER

Another crucial aspect of bike geometry relates to the front wheel, specifically its circumference.

It's a common topic of debate in off-road circles that nothing less than a 21" wheel should be used on dirt, leading many to view ADVs with

19" wheels as less suitable for serious off-roading. Setting aside personal goals and what might be the best wheel for achieving them, I'd like to steer your attention towards a typical misconception in this debate.

The reality is, off-road can be effectively done with any wheel size, as long as you understand what that entails.

A 19" wheel will undoubtedly have a lower axle-to-ground distance compared to a 21" wheel. Yet, either of these wheels can easily roll over any obstacle below their axle line without the need for any special riding techniques.

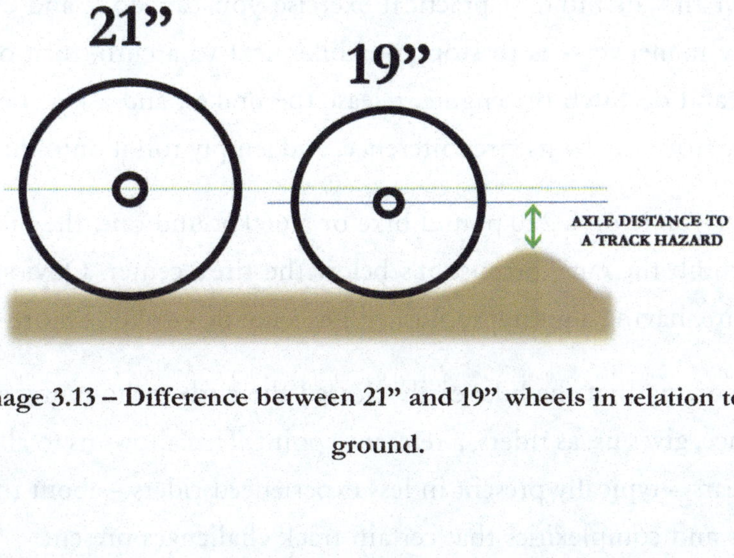

Image 3.13 – Difference between 21" and 19" wheels in relation to the ground.

Ignoring maneuvers like pre-loading the suspension or popping a wheelie, which allow us to completely clear an obstacle or modify how the front hits something, if in our riding we stick to terrains where hazards are below our wheel axle, the size of the wheel doesn't necessarily pose a problem.

Topics like angular velocity, tire contact point rotation speed, and even the impact angles of different-sized wheels on the same obstacle are indeed valid. However, they are more complex to understand visually and mathematically, offering limited practical benefits for this explanation.

The difficulties or potential advantages of rolling with different tire widths on the same terrain - a difference often seen between 21 and 19-inch wheels - are also calculable and debatable. But in terms of direct circumference, physics is clear and its visual understanding direct, offering an immediate benefit in terrain choice for the rider, at any level.

With all this in mind, a practical exercise you can do - and a great recovery maneuver - is to stop your bike next to a curb, turn off the vehicle and declutch the engine, release the brakes, and ask a friend to grab the front tire by its circumference, and simply roll it onto the curb.

Whether you're on a 220-pound bike or a 660-pound one, the bike will easily climb the curb because it's below the tire's center. Obviously, a larger tire, having a greater radius, will execute this exercise more easily.

Still, knowing that the bike easily cleared the curb without any engine assistance, gives us, as riders, a reference point. This allows us to alleviate some fears - typically present in less experienced riders - about the real dangers and complexities that certain track challenges present.

In summary, what's crucial to remember is that there isn't necessarily a better or worse wheel size. Rather, there might be a mismatch between our ambitions and the actual capabilities of the bike we choose to fulfill them.

CHAPTER IV

THE MIND OF THE RIDER

USING EYESIGHT WITH INTENT

There's plenty of chatter about the best bikes, best tires, oils, safety gear, and gadgets, yet we often skip something crucial for real safety and performance on a bike: our mental and hormonal makeup.

Just like staying hydrated directly impacts our capabilities, grasping the fundamentals of our vision, instincts, and brain's stimulus processing can radically alter our approach to riding, as knowledge of what's happening within will allow us to make informed decisions, reducing errors. This means that even when things don't go as planned, having a good understanding of the situation enables us to pick the optimal response.

That being said, when it comes to eyesight, our ability often starts off hampered by a basic assumption: whether we can or cannot see. This position hampers our abilities because this viewpoint is far too simplistic for the complex reality of vision, which entails much more than merely ensuring clarity of sight.

As riders in control of our environment, it's vital to recognize that vision divides itself into two distinct parts: image capture and the processing of those images by the brain.

For simplicity, let's then define three types of visual capture in our eyes: foveal, parafoveal, and peripheral.

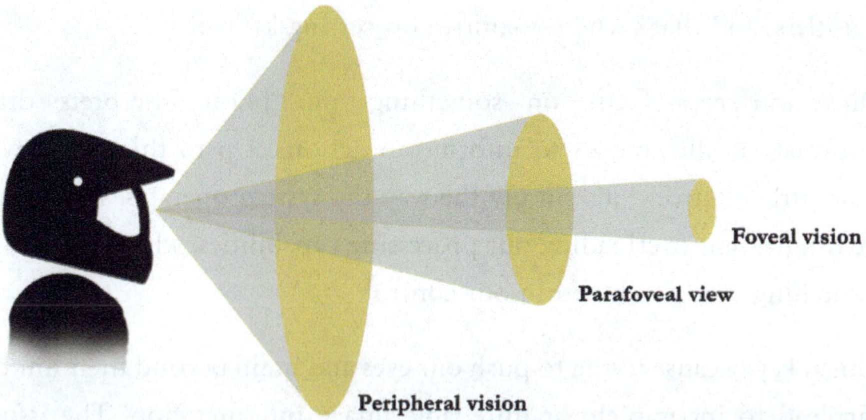

Image 4.1 – Different types of visual capture.

Peripheral vision, the unfocused yet comprehensive view that gives us a broad image of our surroundings, is well-known. However, foveal (maximum focus) and parafoveal (semi-focus) vision are less commonly discussed.

That being said, trying to maximize our foveal vision isn't enough for better sight, as we're limited by our 'hardware' - the capabilities of our eyes and, to an extent, our brains.

That becomes simple to understand if we define that we can't expand the 2-degree span of foveal vision or alter the about 5-degree range of our parafoveal vision on each side. Similarly, we can't change the average 225 milliseconds it takes to focus an image, like a rock.

This focus on different objects is achieved through sudden eye movements called saccadic movements, which allow our eyes to hop between points of interest. Yet, continuously making saccadic movements without processing the information we are seeing is pointless, and that's where cognitive processing kicks in.

Once our eyes fixate on something, our brain interprets this information allowing us to inform our actions. Up to this point, we have little control - it's simply the way the system operates. However, we can prevent overloading our processing capabilities, which is finally something very much within our control.

This is key because trying to push our eyes and brain beyond their limits can lead to incomplete or unreliable image interpretation. The issue arises when working above our capabilities. The brain is forced to fill gaps in terrain reading from a lifetime's database of experiences, a process bound for failure in the long run.

Picture a puzzle with a couple of pieces missing. Although the pieces are absent, we can easily infer their shapes and content, thanks to a mental database of similar situations we've encountered in the past. However, relying solely on this ability doesn't ensure the accuracy of specific details in each image.

For instance, while riding, filling in track segments from memory doesn't guarantee noticing unique elements like an animal, a hole, or other hazards.

This processing overload increases the chance that our perception may not be entirely credible, leading us to make technique or path choices based on sketchy information. This situation creates clear safety and consistency issues in our riding, a familiar scenario for those who by riding faster than their comfort speed found themselves colliding with something they "simply didn't see."

Knowing this, riding within our limits and accepting our skill level gives our brain the necessary time to effectively process information, enhancing its capabilities. Therefore, acknowledging and embracing our current developmental stage, including its limitations, is the first step to surpassing them and maximizing our existing skills and advancing beyond current limitations.

On the flip side, we encounter the issue of prolonged fixation on an object, known as target fixation, or the old saying, "if you stare at the rock, you're going to hit it."

To understand this phenomenon, picture riding and suddenly encountering a hazard, like a rock, rut, sand, or something similar. The term 'hazard' is key because our natural instinct is to perceive the challenge ahead as a danger needing resolution.

This means that unintentionally, our gaze will lock onto this hazard until our brain devises a solution. Practically speaking, our vision fixates on the rock - for instance - prompting our brain to sift through its database for the best way to pass it. Once a solution is found, our eyes will shift to another point of interest, such as a log, while the rock problem and its solution remain queued up until execution time.

Understanding how our brain formulates solutions is then crucial because, once again, we do have some control here.

This control comes by keeping up with our training, enhancing our database of solutions and gaining experience, which broadens the range of information and options our brain can access when faced with a problem. Just think about the myriad of ways you've tackled different rocks on the trails. Generally speaking, technical knowledge about how to navigate rocks, combined with our accrued experience, makes us comfortable when facing yet another rock.

However, occasionally we will encounter a rock that, due to its unique shape or position, captures our attention differently, leading us to fixate on it until our brain decodes a solution. As mentioned, this fixation persists until a solution is found. Sometimes, though, a solution might not exist, either because we've never faced such a problem or lack experience with it, necessitating extra time to devise the best attack plan.

This can be likened to using an app on our phone. If it's an app we use frequently, we find it quickly, often without needing to look at the screen. Conversely, if it's a less frequently used app, it takes longer to locate, and for those "someday I might need this" apps, we might not even remember we have them in the first place.

Returning to our practical riding example during this mental search time, our gaze tends to stay fixed on the issue, be it sand, a rock, a rut, or any other problem we face on the track. This fixation time is critical because focusing on something while riding inadvertently aligns our body with our vision, directing us straight toward the problem or hazard we're fixated on, a danger I believe I won't need to expand on.

This is why the difficulty level of the terrains we ride and the speed we traverse them should increase incrementally, being directly related to our foundational technical knowledge and experience. Adopting this simple approach prevents us from fixating on hazards for too long, as regular training expands our brain's procedural memory, enabling it to swiftly resolve the myriad of challenges we are bound to encounter on the trails.

THE MICRO ADJUSTMENT PROBLEM

Ever found your computer frozen, just because you were writing an email while having 300 tabs open in your browser and running four or five programs simultaneously?

We all know the real issue wasn't the email, but the overload and misuse of the limited resources. On the bike - allowing myself some scientific liberties with this explanation - we can say our brain functions similarly, responding not only to immediate stimuli but also to background processes.

As such, knowing we have processing limits, we're forced to maximize existing capabilities without overtaxing them, avoiding a system crash, a simple yet often overlooked ability.

When we apply this concept to our understanding of vision and procedural memory while riding, we start to see why looking far ahead is generally better than focusing too close.

Looking far allows our brain more time for necessary saccadic movements to choose lines and spot dangers. This leads to the background selection of techniques to move forward, resulting in the best possible plan of attack.

However, due to the bike's design and its geometry, as we've previously analyzed, we also need to give it the freedom to move. By looking far ahead, we not only prevent our 'brain computer' from crashing, but also maintain the composure to let our bike do what it needs to.

Yet, this doesn't mean always looking far ahead is the best choice, as the preferred position for our vision depends on our speed and the type of terrain we're navigating.

We can then say that if we're on a highly technical route that limits our speed, we'll need to focus closer to the front wheel than if we were on an open desert track at full throttle, for instance. This forces us to stay aware that whether we look near or far doesn't change the processes happening in our brain or bike; it just means they operate at different speeds, and we need to respect this reality.

A slightly deeper analyses tells us that traveling at a certain speed and looking far ahead allows our brain to collect, analyze, and pass necessary information to our muscles on autopilot. Conversely, maintaining the same speed in the same terrain but focusing close will overload our systems, as it will feel like the ground is rushing at us too fast, triggering instinctive brain responses that may be unnecessary or even counterproductive.

A clear example is in scenarios where we focus near the front wheel while traveling at speed. Seeing the handlebar turn because we've hit a rock, for instance, will force us to instinctively pull the bars to the other side as quickly as possible to prevent a fall. However, this instinctive micro-adjustment, instead of helping, can completely negate the bike's self-balancing work promoted by the trail, which is what would actually protect us from a fall.

With this in mind, we must accept that it's our responsibility to be mentally present enough to not hinder the bike's functionality or put ourselves in a situation where a mental system crash is inevitable.

At this stage, we can conclusively say that whether we should look near or far should be determined by our traveling speed over a particular type of terrain. Yet, leaving it at that would be oversimplifying and potentially misleading, as whether travelling at low or high speed, we should always scan as much as possible to avoid losing a general sense of the terrain ahead.

This means that if we're traveling at speed, we should mostly look towards the track's limit to choose the best lines, while also constantly making saccadic movements nearby to detect immediate dangers and map out what's ahead. At low speed, the reverse is required. In this case, we should primarily focus our vision closer to the front wheel to choose the best immediate trajectory and dangers to avoid, while continuously scanning the horizon, allowing our brain to choose the best overall path in the background.

Ruts are a typical situation where we're forced to make frequent close saccades but also continuously scan for potential exit points and the "rail" that is leading us there.

With all of this in mind, we're compelled to accept some realities.

First, we're highly dependent on the quality of our base training, as without it, our brain won't have sufficient information to automatically and directly choose the best solutions for the problems we encounter, whether they're rocks, ruts, or sand.

Alternatively, we can acquire this information through experience on the bike, but even here, the lack of good foundational training can dictate that the information gathered is underutilized, either due to a lack of context or a careful analysis of its true meaning.

We also have to accept that if we don't change our attitude with the type of terrain and speed, our brain will tend to lead us into mistakes, such as focusing too long on dangers, having terrain reading failures, or instinctively making micro-adjustments that contradict the bike's natural work.

Therefore, by understanding how our vision directly impacts our reactions and how those implications can be either a benefit or a risk, we can more easily comprehend our limits and adjust our safety envelope to our real knowledge and technical experience.

BREATHING CONTROL

Now that we've grasped why we should pay attention to what lies ahead and why, it's equally critical to understand the importance of our breath while riding off-road.

On one hand, we must remember that riding a dirt bike - be it an enduro, MX, or ADV - is a sport, regardless of how intensely we pursue it. On the other hand, we must recognize the significance of breathing in the human body.

I know this last point may seem obvious; after all, we all know that without breathing, we'd perish. However, there's a facet of how we breathe and how our bodies react to stress on a bike that eludes many riders, from weekend warriors to professionals.

Whenever we find ourselves on the bike in a discomforting situation that demands prolonged and heightened alertness - like a hazard, particularly challenging terrain, or a maneuver or speed beyond our capabilities - our bodies trigger a physiological response. For most people, this response means going into apnea or, in other words, holding our breath.

This bodily response can be found in various situations, from a suspenseful sequence in a movie to the overwhelming feeling of seeing 1000 unread emails in our inbox at the start of the day. Instinctively, we tend to hold our breath.

In our world of two wheels, it's not uncommon to hear professional motocross and supercross athletes recounting how they sometimes go half a lap or even a full lap on the track without taking a breath. For us, recreational riders, it's easy to experience this whenever we venture out of our comfort zone, and that's the crucial point to bear in mind - our comfort zone.

There aren't many fears that are common to all humans, but the buildup of carbon dioxide in the blood is undoubtedly one of them. Building tolerance to CO2 is unquestionably possible, which is precisely what free divers do. But that's not our focus here, nor should it be.

What we need to understand is that our bodies don't have an oxygen meter; instead, we have neurons that sound the alarm whenever our carbon dioxide levels rise in the blood, compelling us to take rapid, sigh-like breaths as a way to balance the levels of CO2 and oxygen in our bodies.

This alarm, combined with the hyper-focus that initially led us to hold our breath - as a reaction to being out of our comfort zone - triggers an even greater excitation of our autonomic nervous system. This causes our pupils to dilate, creating tunnel vision, reducing our peripheral vision, and also initiating a surge of adrenaline and cortisol that raises our heart rate and blood pressure, leaving our muscles tense, among other bodily responses.

These are typical symptoms of a fight, flight, freeze, fawn, or forget reaction.

This reaction, present in all of us, serves as a way to respond to stress, fear, and dangers, and it's entirely instinctive, forcing our bodies to react before our minds can calmly analyze the situation, as the frontal cortex goes into automatic mode, bypassing our analytical capacity during this process.

This inability to control our reactions is much more noticeable in people with less training, in any situation. It is therefore one of the main reasons why training greatly enhances riders' safety.

Training, however, isn't just about repetitive exercises; it also involves explaining the existence of these and other phenomena, as our imagination is so powerful that, even in a completely safe situation, if we convince ourselves that something will go wrong, we'll force our bodies to have the exact same reaction they would in a real danger situation.

This reality alone carries clear risks on a bike, and it gets worse when we consider that the physical responses of fight or flight take the body between 20 minutes and several hours to fully recover once they've ended.

With an understanding of theoretical explanations, we can navigate preconceived fears, significantly reducing the likelihood of putting ourselves in dangerous situations unconsciously. With repetitive exercises of increasing difficulty, we make it possible to quickly and safely push the boundaries of our comfort zone.

Military personnel, emergency teams, combat sports athletes, among others, for instance, are accustomed to dealing with this type of physical response while maintaining high cognitive abilities. But, as we mentioned, it's not because they're different beings from any of us; it's because they've achieved this presence of mind through extensive training, whether through simulations or real situations.

In this way, by combining theory and progressive practice, we allow ourselves to have the mental capacity to relax, focus on our breathing, and let the brain apply the procedural memory it has created or is creating.

That being said, it may seem somewhat counterintuitive to focus on our breathing rather than the exercise or difficulty of the task at hand, especially when we're learning. However, the way we breathe is the key we need to unlock certain mental blockages.

This becomes a full blown reality as by going into apnea, the body's primary focus is not to execute the exercise with clarity and precision, but rather to finish it as quickly as possible, in order to return to our comfort zone and reactivate basic systems, like normal breathing.

Therefore, the only thing we need to keep in mind, especially when we feel that we are entering - or about to enter - a situation outside our comfort zone, is to focus on our breathing.

Nevertheless, simply breathing won't be the greatest help we can give ourselves.

What we want to do is reduce our respiratory rate, focusing on slightly deeper breaths than we would naturally take.

This careful breathing will allow our brain to concentrate on a natural movement, lowering our stress levels, enabling the theoretical training we've had to come more naturally, and helping us to complete the movement or overcome the upcoming challenge more easily.

A typical example where this type of problem arises is when we're learning low speed turns, where similarly to those taken on the road, where we have to lean the bike, these slow turns sometimes require fast and steep vehicle inclinations.

As human beings, we are highly visual creatures, so when we see our bike approaching the ground, we tend to assume we are losing the battle with gravity and are about to fall. This phenomenon occurs whenever there is an abrupt change in visual depth perception relative to the ground; and it's entirely instinctual.

Just imagine a newborn baby with barely open eyes sensing it might be falling and, immediately, opening its eyes in response to the stressful situation.

As adults, we can experience this when we feel nervous and have reactions like sweaty hands or an accelerated heart rate while watching videos of free climbing or roller coasters, for example.

So, returning to the example of learning low speed turns, for many of us, when we see the bike falling and don't focus on our breathing as a way to prevent an exponential increase in our stress levels, we quickly forget our knowledge of physics and riding technique and let our instincts take over, effectively increasing the likelihood of the bike falling, by botching our usage of the clutch, brakes, or body position.

Once again, we demonstrate why knowledge, in addition to riding techniques, is crucial for overcoming certain fears and ensuring a consistent and safe progression in skill level.

Likewise, this explains why learning at low speeds is essential, as even if we fall during a slow turn, where the risk of injury is lower, the brain begins to perceive that maneuver differently.

The same fall in a similar turn, but at a higher speed, on the other hand, increases the likelihood of injury or damage to the bike, confirming to the brain that its fears are real, which compels an adverse reaction to the movement in subsequent attempts.

This is why it's possible to argue that riders who train at low speeds or come from disciplines in our sport performed at lower speeds, such as trials, tend to have greater ease when riding in faster categories like enduro or ADV. They aren't unconsciously battling deeply ingrained fears like other riders who, only training at high speeds, over time, see the confirmation of their fears having very real consequences.

For all these reasons, I would say that perhaps the greatest asset we can have on the bike is to focus on our breathing, because that's what will give us the peace of mind we need to consistently and confidently read the terrain, apply our best technique, and stay focused on the signals our bike and our body are giving us.

CHAPTER V

KNOW YOUR BIKE

No one debates that to be able to ride a bike, you need to be on top of it. Even though some take this definition to the limit with tricks worthy of any circus; they tend to be, in general terms, on top of the bike.

That said, I believe that in order to make the most of our adventures, knowing how to handle the bike off of it is a fundamental technical skill, which requires us to talk about it before anything else.

For many, going to a motorcycle dealership or workshop is an activity limited to seeing machines and gear, or simply repairing or installing parts and components.

For me, even though that reality is somewhat similar, there's something else that always captures my attention, which is seeing the mechanics and salespeople moving bikes from one place to another.

The lightness, ease, and almost disdain with which bikes are moved with precision, whether they're on or off the seat, is, at the very least, hypnotizing.

On the other hand, looking at most motorcyclists moving their bikes, whether in garages, getting in or out of vans, or in various parking lots and trails, is a total contrast to the previous reality.

The reason for this dichotomy is relatively easy to understand: while some spend hours of their day training on how to move bikes in the fastest and easiest way, others choose not to.

That said, and considering that most of us are not comfortable or at ease moving our bikes when we are off of them, should we learn something from those who do it totally nonchalantly?

Without a doubt!

After all, not all off-road riding situations involve being on top of the bike to start with.

Image 5.1 - Water crossing with the rider next to the bike.

This is one of the reasons why it is very common in ADV bike training to start with getting acquainted with the bike, whether through static exercises with the bike turned off or dynamic ones with the bike in motion, but not necessarily with us on top of the bike.

However, this type of work is not as common with smaller enduro bikes, and as far as I'm concerned that's a mistake.

In ADV bikes, this type of exercise has a clear purpose: not only to introduce the user to the dimensions of the bike but also its balance point and weight distribution.

When we understand that a 500 pound bike can be controlled with one hand, or even a finger, our perception of size, weight, mass distribution, and the impact we have on the bike changes completely. Therefore, this learning step is just as valid and crucial for a 550 pound mountain as it is for a 220 pound lightweight.

With that in mind, if in the "big" one, having this knowledge deeply ingrained dramatically changes how we approach the bike, in an enduro bike, it changes the types of terrains where we will feel comfortable, for example.

Nevertheless, the benefits don't stop there. If there's a common stressful situation for almost all off-road riders, it's the act of being stationary, by choice or obligation, in a place where we are forced to stay off balance.

Imagine a place where we were forced to stop or even dismount, next to a ravine, on a steep climb, or even after a fall, and we find ourselves without a good foothold or in a good position to get back on the bike. Knowing how to handle the bike when we are off of it opens up a world of possibilities, allowing us to remain calm and, with controlled breathing and heart rate, make the best decisions, even in situations like the ones in our examples.

Another advantage is that learning to handle the bike where we are off of it requires little more than just the bike itself, with many of the exercises even being possible to achieve with minimal safety equipment, in the comfort of our own garage. As we progress in the exercises, we will obviously need the rest of the equipment and to look for larger locations such as parking lots or clearings, but to start, it's important to allow ourselves to relax, so we don't leave our comfort zone so abruptly and can deeply focus on the task at hand.

It's also important not to forget that repetition is key to memorization, so introducing some of these techniques and exercises into our warm-up before rides can also be an option to consider to maintain proficiency and start the day with small goals achieved, something that will drastically change the attitude with which we will face the trail, by setting the brain and body chemistry deeply in our favor.

The list of exercises and possible variations - is immense - even in my training with students I can’t use them all - so creating a encyclopedia-like list here would be more confusing than anything else. However, I couldn't address the topic without sharing at least some exercises, as well as their purpose, thus creating a foundation that will allow you to develop your own routines and variations.

GENERAL MEASUREMENTS

The first drill I recommend happens with your bike turned off, but in gear, and not resting on side and/or central stand. Now move slowly in a full circle around it, keeping just one hand on the bike at any given time.

Image 5.2 - This exercise should be practiced in both directions.

Starting on the side of the sidestand, where most of us feel more comfortable, the first step is to, with both hands on the bike, remove the stand and find the bike's balance point. By balance point, I mean the spot where the bike stays upright without any effort on our part, for reference, on a windless day, the bike should be able to balance itself without any input once the balance point is found.

From there, pick a direction to rotate and start your movement, passing the bike from hand to hand until you get back to where you started. It doesn't really matter where you grab the bike, as long as it's a structurally sound part - which, by definition, a turn signal isn't.

I advise doing this exercise in both directions and recommend that, between each hand swap while holding the bike, you take the time to re-find the bike's balance point before moving on. Remember, the goal here is to build confidence and comfort, not to rush through the exercise, as rushing will only work against you.

Mastering this simple technique will be the first step into making you comfortable with your bike in all its entirety, allowing you, whenever you're off the saddle, to stay relaxed no matter your position relative to the bike.

TOO MUCH OF A GOOD THING

A big issue when we talk about the weight of bikes isn't necessarily the weight itself, but where the center of gravity is located, something many mention, but not all fully understand.

Take for instance, slowly lowering to the ground a R1250GSA with its boxer engine acting as a low-slung counterweight, or a 450 Rally with its high tanks and instruments. It might surprise many to find out which one feels heavier at the same tilt angle.

I'm not saying that a GSA is lighter than a 450 Rally, by any means, but rather that their weight isn't everything, especially when you're moving away from a completely upright position.

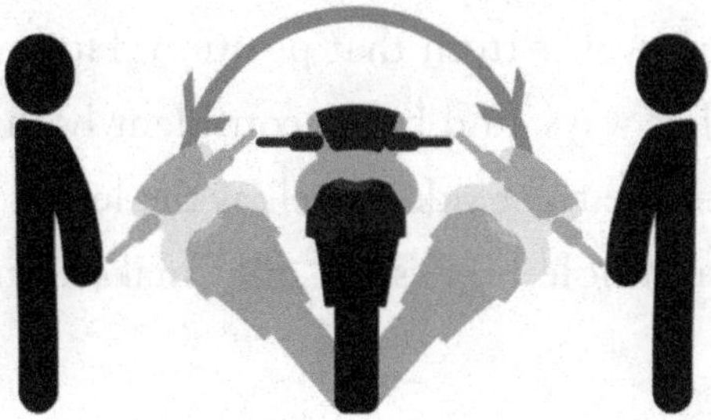

Image 5.3 - It's crucial to understand how far we can manage to hold a tilted bike.

This means that, once again, with the bike turned off and in gear, slowly lowering it as if you wanted to lay it on the ground will clearly indicate the maximum point you can go with that specific bike before it becomes too heavy for you.

This knowledge will not only unlock some dynamic techniques but also help us decide when it's a good time to abort a maneuver and let the bike fall, instead of trying to force a potentially injury causing save. For example, trying to put your foot down to save a falling bike is how many avoidable injuries happen.

Knowing in advance where that limit is and what it feels like - something our brain records and later uses - allows us to avoid problems more effectively, and increases the comfort we have with our machine of choice.

EVERY BIKE HAS TWO SIDES

By default, we all learn to get on and off the bike from its left side, just as we tend to maneuver it from that position. However, every bike can be handled in other ways, and being confident both in mounting and maneuvering from the right side as well as the left, is a basic skill. After all, we pay for the whole bike, so it only makes sense to use it in its entirety.

Thus, developing this technique is as simple as it seems. Just use the knowledge from the first exercise, turn your body to the right side of the bike, and learn to maneuver it from there. I recommend, for a full and gradual evolution, that this exercise is done in four stages.

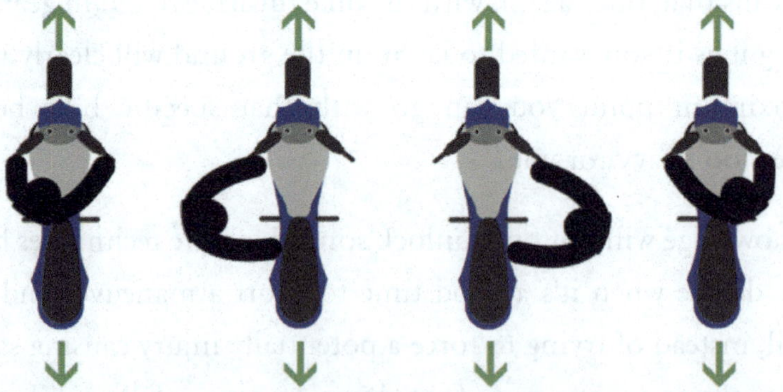

Image 5.4 - This exercise should be performed from both sides of the bike.

1st - With the bike turned off and with you on its left side, take a few steps forward and back, pulling and pushing the bike by the handlebars.

In another variation, instead of pushing, you can move by starting the bike and using its engine, with the aid of the front brake, clutch, and throttle to move forward. For moving backward, initiate the bike's movement by compressing the front suspension.

Contrary to what many do – and deeply believe - using the front suspension to move a bike means actually compressing it downward, and using the stored energy you just put on the springs to initiate the intended movement backwards. The best way to understand this concept is, with the front brake pressed throughout the exercise, trying to compress the suspension by pushing the handlebars towards the ground and observe the bike's reaction.

By doing this movement on its own a few times, you'll start to understand at which point the suspension returns to its resting point, and it is at this point that, if you release the brake, the energy being released by the suspension will translate itself into a movement of the bike backward. Releasing the brake before or after this point will not maximize the maneuver's effectiveness, reducing or completely eliminating the bike's overall motion. Similarly, not compressing the suspension fully downward will limit the energy it has available for use.

This exercise, like the vast majority of them, should not be rushed, and we should give ourselves time for each movement of the body and bike to do what they need to do before attempting the next step.

2nd - With the bike turned off, you can put one hand on the handlebar and one on the rear of the bike, either on the seat or a handle, until you find the bike's balance point. At that point, try to take a few steps backward and forward, pulling and pushing the bike by the handlebar and wherever you've chosen to place your right hand.

Increasing the difficulty of the exercise, you can start the bike and take a few steps forward, using only fine clutch control, and a few steps back, disengaging the bike by engaging the clutch, and pulling the bike by the handlebar and the seat or handle, keeping it always in its vertical balance point to the best of your ability.

3rd and 4th - Using the attached image as a reference, it's easy to understand that the next two exercises are a repetition of the first two, but this time, done from the right side of the bike.

The great difficulty of this variation is the lack of habit that most have in using the bike from the right side, making these exercises an excellent learning step.

As a note, I don't want to close these exercises without talking about the training of what is, in reality, the best way to move a bike in most situations, especially off-road.

To achieve that, we should create a V shape between ourselves and our bike, leaning it against our hip/trunk in a position where we are comfortable and relaxed with the bike tilted. If you are not comfortable enough to remove your hands from the bike once the position is achieved, and have a conversation while being stationary with the bike, you are not in the right spot.

From this balanced position, any small movement of your hip will make the bike move forward or backwards, without the need for you to push it by the handlebars - or use the engine -, removing instability from the bike, and reducing physical wear.

With all this in mind, being able to use the bike from all sides with equal comfort and confidence, as well as having the guarantee that we can do it in the most varied ways will unlock many fears that later appear in dynamic maneuvers, as well as will change the way we see our bike as a whole.

As a final note, for those with DCT bikes, or Rekluse semi-automatic systems, you can make your forward movement with the engine using the throttle and front brake, which is a great method to start developing the essential fine control of these commands.

CHAPTER VI

BODY POSITIONING FROM HEAD TO TOE

Having broadly analyzed all the basic points and nuances needed for a deeper understanding of our sport, our bikes, and how our body functions when riding, we're nearly ready to sit on the bike. However, we shouldn't do so without first understanding some concepts about body positioning.

These notions begin with the understanding that any bike would work wonderfully if we could control it remotely, meaning that our need to adopt a good body position arises from the simple necessity to avoid disturbing the bike.

In other words, our job as riders on a bike boils down to two points: choosing lines and general trajectory, and adopting the most suitable body position to give the bike the best working conditions at any given moment. I repeat, our job is to give the bike the best conditions to do what it needs to do.

This unwavering reality forces us to accept that the best and most advanced body positioning techniques appear in the most demanding off-road disciplines, where it's crucial for riders not to hinder the development work done by years of accumulated engineering knowledge. That's why the best professional athletes are those who can make the most of the machine, and not necessarily the most flamboyant.

To maximize our riding capability, we're then obliged to borrow techniques used in the world of enduro, supercross, and motocross, because it's in these realms that off-road bikes, in their purest design, are pushed to their limits.

I say off-road bikes in their purest design because the ADV bikes we use - and even some dual-sports - are compromises, as they must perform well both on and off the road. Enduro, MX (motocross), or SX (supercross) bikes, on the other hand, were designed solely with off-road in mind, a startling difference if you really think about it.

Considering that both in ADV and recreational enduro, especially in adventure settings, the idea is to cover as many miles as possible, it would be easy to assume that we should be borrowing our techniques from Rally Raid, but that would be a mistake.

Returning to the premise that the best techniques were created for bikes in their purest and most simplistic state, when we talk about bikes with some type of modifications, even factory-made ones, we are necessarily referring to an adaptation of the technique to the compromises these changes have created in the bikes, as these compromises generate limitations in the use of pure technique. The extent of these compromises is debatable, and in many cases, somewhat insignificant, but compromises just the same.

Therefore, if we look for inspiration in Rally Raids, we are observing techniques already adapted to bikes with heavier weights, navigation towers that limit movements and visibility, wider fairings, and thus, sometimes altered riding positions compared to an SX or MX bike. This means we would be limiting ourselves by choosing these techniques as a base.

With that in mind, we should then look at the purest technique and adapt it to the limitations of our body in relation to size, physical capacity, and potential injuries, as well as to the limitations of the bike we choose. This logical sequence is what will allow us to ride any bike with the maximum consistency and coherence possible, something impossible to do if our base knowledge comes from an already adapted technique, regardless of the extent of those adaptations and knowledge.

Another crucial notion is that, in order to maximize the performance of our bike - a machine carefully designed to the fraction of an inch - we, as riders, must avoid abrupt and exaggerated body movements in the search for the best momentary body position.

This is the notion that will allow us to reduce our physical wear and tear by eliminating unnecessary movements, while enabling us to work with what the bike is asking of us, instead of trying to forcibly make the vehicle do something that may be difficult for it, or even impossible.

This analogy may seem somewhat esoteric, but there is a very clear dance component between rider and bike, much like when we're dancing with a partner, where both have a clear role to play and an obligation not only to give each other the best conditions to move with ease but also to let the movements being made to be completed in their entirety before starting the next one.

A clear example of this analogy in real life is if, right after hitting a hole, while the suspension is still recovering, we immediately try to force the bike into a tight low-speed turn. Our lack of patience in not allowing the suspension to run its course and the bike to balance itself will drastically change the difficulty of the maneuver, as it creates an instability in the bike that will make the execution of the movement much more difficult, if not impossible in certain scenarios.

Finally, we must accept the notion of a base body position, a concept that arises from the need to always return to it as quickly as possible at the end of a technical movement in an attempt to break a possible chain of error. The concept of an error chain is one widely used in aviation and defines that an accident or incident never happens for one single reason, but due to a chain of errors that, when aligned, can produce disastrous effects.

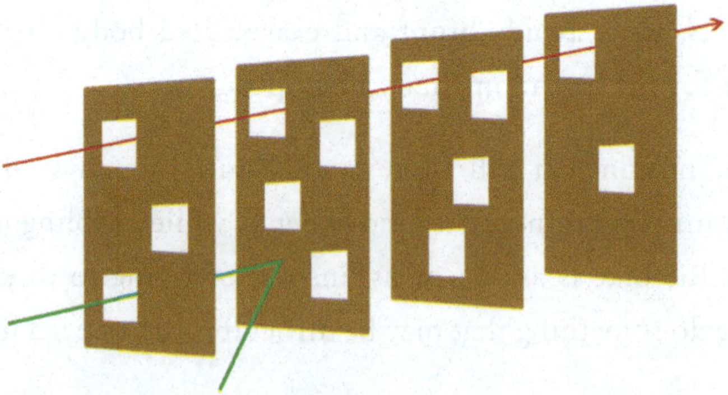

Image 6.1 - An example of the error chain in action.

Thus, our central base position should serve as a method to break any errors that we may have started during a technical movement. Regardless of whether we've made a good technical application or a less capable one, by returning to our initial position after its execution, we will start the next movement from zero, that is, without any major errors accounted for.

In other words, we should always start any technical movement from our base position, returning to it as soon as possible, just like a yo-yo, which, in a simplistic way, does what it has to do and always returns to our hand before the next rotation.

All that being said, there shouldn't be any doubt that, in some cases, it's impossible to immediately revert to a base position, as happens, for example, in sequences of climbing, descending, and interlinked turns. However, this only tells us that we should make these body position transitions as smoothly as possible and, as soon as conditions allow, return to the base position to break any errors we may have committed during the sequence of movements we've just executed.

With all these notions present and well understood, it's then time to define what the base position should be, and for that, borrowing a concept from the great Ryan Hughes, let's analyze the human body as a tree.

I choose this analogy because, like in a tree, it brilliantly exemplifies how we can see our head and arms as branches, in the sense that they can sway more freely without destabilizing the tree as a whole.

Image 6.2 – The body as a tree.

Between our waist and chest lies the upper part of the trunk, from where the branches come out, a zone by definition strong and secure, but capable of some mobility, a vital point to prevent the tree from breaking.

Our legs assume the position of the lower part of the trunk, a very strong, very secure area that only has the minimum mobility necessary for the upper part of the trunk to be mobile in its own right. And finally, the feet - the point where we will start our analysis - which thus take on the important task of being the roots of our tree.

We'll start here because, no matter how knowledgeable we are or not about plants and trees, I believe we all accept that by cutting or damaging a root, it's difficult to keep a plant healthy. The same goes for us, a bad foot position has serious implications, so it becomes important to start not only by defining what a good foot position is and why but also to diagnose some typical mistakes and their consequences.

FEET

We've established that off-road riding is a sport, regardless of the pace at which it's done. We've also stipulated that this is a highly physical sport, meaning each of our movements should be intentional, allowing us to conserve energy and avoid destabilizing the bike. Lastly, we've determined that our main task on the bike is to provide it with the best conditions to operate at its full capacity.

With all this in mind, and seeing our body as a tree, in the case of our feet, we're talking about a living, moving root - essential for maintaining the stability of the rest of the body. If our feet stay in one static position throughout our ride, we'll never be able to maintain or solidly return to our base body position. Thus, we will never effectively maximize our efforts or reduce our body error chain to zero. All of this means that if we're not moving our feet while riding, we've got a fundamental issue with our body technique that needs quick correction to enhance our safety and consistency.

A common foot-related problem is often referred to as "duck feet", which contrary to popular belief, doesn't stem from one error alone but two, both deriving from an intuitive view of a counterintuitive sport.

A quick look at boots with heels, commonly found in our sport, gives the untraining eye the impression that we should anchor the heel of the sole on the footpeg, a position that places the foot much further forward than it should be. This forces the toes to rest on the controls, naturally causing us to point our feet outward to avoid accidental contact with the brake or gear shifter, resulting in the typical "duck feet" position.

This incorrect position becomes a constant risk for injury because it not only makes our feet the widest part of the bike at that level, it also creates an additional point of contact with our surroundings, as when the foot turns outward, any stone, branch, or even the ground upon contact with the boot will not rotate around the toe box. In fact, any contact will try to force the foot off the footpeg, putting the body in a precarious safety position. This body rotation, in addition to forcing us to break contact with the bike, creates the base conditions for a foot-to-groin rotation that can cause severe ligament and muscle problems.

Image 6.4 – Example of good and bad foot positioning on the footpeg with a frontal impact on the foot.

Similarly, when our only security to avoid accidentally touching the controls is to rotate the foot outward, any distraction or situation where it's impossible to turn the foot out, like riding in a rut, exponentially increases the likelihood of accidentally shifting gears or using the rear brake.

That being said, to increase our safety and technical consistency, especially as beginner or intermediate riders, we should strive to keep our feet back on the footpegs, resting in the area where the toes begin. This retracted position has several benefits; it not only hides and thus better protects the foot from an impact that could dislodge it from the footpeg but also requires us to calculate every foot movement, forcing a deliberate and precise motion every single time.

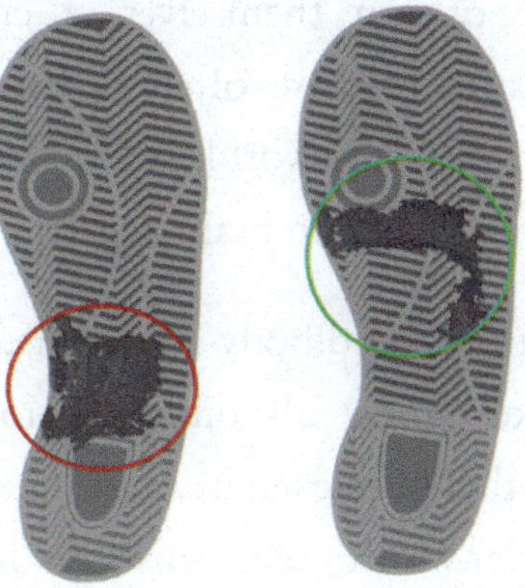

Image 6.3 – The difference in sole wear of boots between good and bad technique.

It's arguable that moving the foot back and forth to use the controls is slower than simply rotating the toe from outside to inside. But when the overall benefits greatly outweigh this potential disadvantage, choosing the right foot position becomes obvious, from leisure riding to the highest levels of professional sport.

Going back to our tree analogy, just as the roots support the start of the trunk, our feet should align our legs. To do this in the most natural way, we need to make another minor adjustment to our feet, allowing us to maximize our biomechanical benefits. I said minor, but we must keep present that in a vehicle that is designed to the fraction of an inch, minor adjustments have major impacts, so our position and movements should be no different and should always be seen under that lens.

Therefore, foot placement should be done by pulling the foot back and slightly turning the toe towards the bike frame. This position, combined with a squat - a typical body movement on the bike - changes how the knees automatically position themselves, a crucial point for bike control. This automatic response often doesn't get the attention it deserves in our two-wheeled world, which is a mistake, as any automatism we can create frees up brain capacity as well as saves energy.

The quickest way to understand body automatisms is through practical exercises off the bike, where all our attention is focused on the movement itself. For this knee exercise, all we need to do is stand with our feet about shoulder-width apart, turn our feet out, squat slightly, and try to bring our knees together. Unless someone has unusually flexible joints, trying to touch knees together in this position will present itself as extreme difficulty, or as an overall totally impossible maneuver.

On the other hand, repeating the same exercise with feet parallel greatly simplifies the ease of closing the knees, and if repeated with feet slightly turned inward, it becomes clear that no effort or mental capacity is needed to bring the knees together, as they naturally and automatically touch.

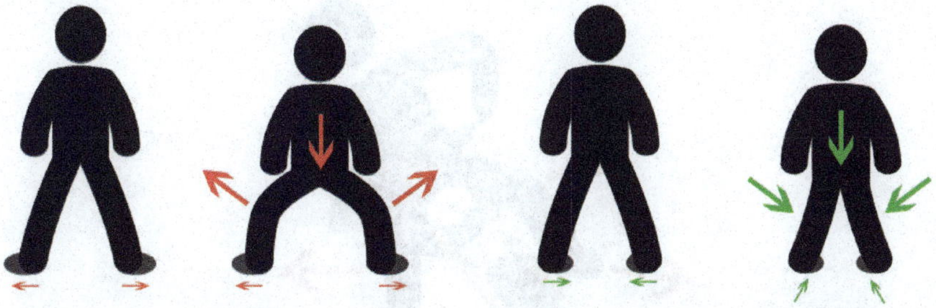

Image 6.5 – Knee automatism.

This automatism ensures constant, effortless contact between the knees and the bike - whether sitting or standing - becoming a key position in promoting bike control. However, this control gained on straightaways is lost if we apply the popular maxim that we should remove our foot from the footpeg to turn, a maneuver that usually - except in specific situations - tends to be regularly dangerous and counterproductive.

As demonstrated in the squatting exercises with various foot positions, the difficulty - or even impossibility - of closing the knees with feet turned outward was immense. This tells us that removing our feet from the pegs while cornering - or in any other situation - will make it very hard if not impossible for us to remain closed in with the bike.

Returning to the tree analogy, if we cut the trunk, the tree will tend to fall rather than remain stable.

Image 6.6 – Poor foot positioning in a turn.

Therefore, removing a foot from the peg results in a total disconnection from the bike, stripping away our ability to control it, leaving us attached to the bike only by the seam of our pants and some faith.

However, this doesn't mean that removing the foot from the peg should always be avoided; it simply means that just as we don't refuel the bike every other mile just because we can, this technique should only be used when strictly necessary.

With this in mind, to increase our safety and technical consistency, especially as beginner or intermediate riders, the need to remove a foot from the pegs is reduced to only two occasions: a "dab," or touching the foot to the ground to rebalance the body; and when the foot must leave the peg due to the risk of aggressive ground contact.

If in a dab the foot should return to the peg as quickly as possible to regain control of the bike, in a turn where the foot must leave due to risk of ground contact, it should not be simply removed willy-nilly. In this case, the foot should be repositioned as high as possible, while still keeping a rotation of the toes inward towards the front fender, touching it if flexibility and the bike's fairings permit it.

This is because, if the foot is removed to avoid ground contact, the only logical position for it should be as far away as possible from the danger that forced it to move out in the first place. This logical sequence is impossible to maintain when the foot leaves the peg to be pointed in the direction of the turn, leaving the foot lower than the peg itself - a typical movement for many beginner to professional riders that leads many more to perpetuate this mistake.

Thus, by raising the leg high and tight and rotating the toes inward, we maintain the knee pressure automatism against the bike, ensuring that even in this physically difficult position, we keep our contact and control of the vehicle. Likewise, on bikes that allow it, resting the leg on the side plastic provides extra leg support, greatly reducing the physical strain of the maneuver, making it easier to execute.

Image 6.7 – Good foot positioning in extreme turns.

As riders creating mental procedures relating to our body position, by avoiding removing our feet from the pegs just for the sake of it, we force ourselves to create the good habit of keeping them on the bike as much as possible. This simplifies our turning maneuvers, increasing our consistency in this movement, as we more easily can tackle all turns following the same basic checklist.

These are the main reasons why more and more top athletes, like Jett Lawrence, Chase Sexton, Ken Roczen, AJ Catanzaro and others, are choosing to work on their technique in a way that keeps their feet on the pegs throughout the race. This includes not removing their feet from the pegs even in situations like square turns, with some pushing this limit even in ruts deep enough for their feet to be dragging dirt.

Is it always possible to keep our feet on the bike in the heat of competition, or in the natural relaxation of a Sunday ride with friends? No, of course not, but that's where the benefits of using a simplistic, repetitive maneuver with automatisms and low cognitive impact come into play. A foot leaving the footpeg, for whatever reason, sets off an alarm, but because we still have available cognitive abilities, it allows us to remain calm and capable of returning to our base riding position at the end of the movement, nullifying any previous errors.

However, foot positioning issues don't end there.

Nobody questions that suspensions with about 12 inches of travel are something extraordinary, but when we're talking about a sport measured in fractions of inches and seconds, everything makes a difference. By using the front of our foot on the peg to better control the bike and increase our safety, we're gaining time, as well as extra suspension.

In terms of time, this refers to the milliseconds needed to transition from a seated to a standing position, which may seem irrelevant in leisure riding but is actually a safety factor. When accidents and incidents happen in the blink of an eye, a millisecond for a position transition that doesn't require many adjustments can make the difference between having the ability to react, or not.

As for the extra suspension, this point requires us to look at our biomechanics on the bike, and the best way to open this door is with a simple practical exercise.

If we stand up and try to jump, it's clear we'll tend to land on the balls of our feet first, using the ankle as a shock absorber, though in larger jumps, the knees will also bend after the ankles have exhausted their range of motion.

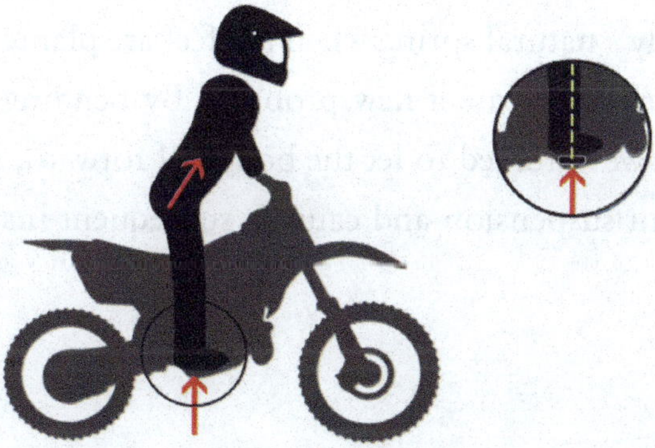

Image 6.8 – The impact on the spine promoted by poor foot positioning on the footpeg.

By landing with the whole foot planted - which our little exercise showed we don't naturally do - not only will we tend to lose balance, especially in larger impacts, but we'll also feel the impact more throughout the entire body, as a basic rule of physics says that energy doesn't get lost, it is transferred until it dissipates. So, with our foot poorly positioned on the peg – that is, on the heel or the central part of the foot - we're eliminating the possibility of using the ankle as a shock absorber, forcing the energy from impacts to be transmitted more violently to the knees and spine. This is one reason why many riders, from professionals to leisure riders, complain of back pain in off-road riding.

Knowing that there's a clear need to absorb impacts, many riders, either out of self-preservation instinct or having been warned of the dangers of spinal impacts, then tend to try to absorb impacts with their knees. Unfortunately, this solution is far from ideal.

On one hand, we continue to transmit a lot of energy to the spine by limiting the body's natural spring when the feet are planted on the pegs, and on the other, we create a new problem. By bending our knees to absorb impacts, we're forced to let the body fall forward, increasing the load on the front suspension and causing subsequent instability in the bike.

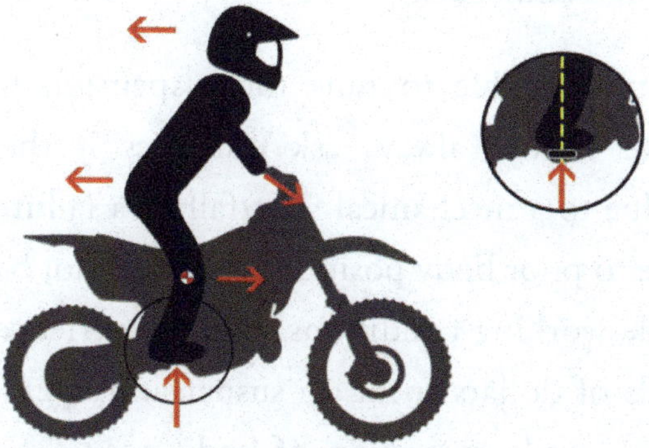

Image 6.9 – The result of using the foot as a pivot point in impacts, pulling the body back and forth.

In isolated jumps, this can somehow be compensated with an increase in effort and physical instability, but in situations of consecutive bumps, like a section of whoops on a track or trail, things tend to change. By sinking the front too much due to poor body positioning, the natural rebound of the bike will want to throw our body back, which will mean that on the next impact, we will be forced forward again, sinking the front suspension further and so on, leaving our body in an accelerating pendulum effect.

Similarly, this effect opens the door to a suspension effect called packing, defined by sequential suspension loads without giving it time to return to its base position. In practical terms, if we have a suspension with 8 inches of travel, and we exhaust it in an impact and don't give it time to recover, in the next impact, instead of our normal 8 inches, we might only have 6 inches available, for example. If this sequence continues, we'll reach a point where we simply won't have any suspension travel available, and there will be an impact that will just throw us over the handlebars.

Undoubtedly, it's possible to tune the suspension to recover more quickly, but we should always ask ourselves if the need for this adjustment is due to a mechanical shortfall or a failure on our part as riders, who, due to poor body positioning, might not be giving the bike the best possible working conditions. From experience, I can say that many thousands of dollars spent on suspension upgrades could easily be saved with a simple correction of body positioning, turning that upgrade into a technical makeup effect rather than a clear, effective mechanical improvement.

That being said, besides the packing problem, by absorbing impacts with our knees, as we are thrown back and forth on the bike like a rag doll, our natural tendency will be to grab the handlebars for dear life, opening the door to problems known to many as whiskey throttle and/or arm pump, two important phenomena to understand, as we should avoid them at all costs.

Image 6.10 – How incorrect knee positioning promotes the body's fall forward of the bike's perpendicular line to the ground.

Whiskey throttle is then the name given to the result of an automatic body movement in a situation of imbalance, associated with the operation of the bike's throttle. Going back, thinking about our body being thrown like a rag doll due to imbalance, the natural reaction of the body will be to grab onto whatever we can, in this case, the handlebars, but on the bars, the throttle twists.

Thus, when we are in imbalance backward and try to balance ourselves by grabbing the throttle with our right hand, we inadvertently accelerate the bike further, which, in physical terms, propels the bike forward while pulling our body back even more, increasing the imbalance and inadvertently forcing us to accelerate even more, creating a growing effect of "firing" the body to the rear of the bike. It is undoubtedly possible to break this cycle, either by pulling the clutch, which effectively cuts power to the wheel, or by having the knee automatism in operation, something that will allow our tree to support itself by the trunk and not the branches. In other words, by committing an error that starts a whiskey throttle, because we can support the weight of the body with the knees, we can free the hand from the throttle, and break the error chain that would inevitably lead us to the ground. That being said, once the whiskey throttle kicks into high gear, it is virtually impossible to recover by readjusting the body, making prevention a key safety feature.

Regarding arm pump, we don't necessarily need a dynamic body imbalance for this phenomena to appear, it's enough that once again, we are not holding our body with our legs, and we are supporting our weight with our arms. Urban legends and hearsay suggest that arm pump is forearm cramps, but this definition couldn't be further from the truth. So, to be clear on this point, we first need to define what cramps are.

For many years associated with the decline of minerals like calcium, potassium, and magnesium in the body - in other words, dehydration - today we know that although this decline is part of the problem, it's not the main reason why our muscles lock up or twitch uncontrollably. Study after study has proven that the main problem has a neuromuscular root, more precisely, a set of circumstances of wear or predisposition that allow hyperexcitation of alpha motor neurons, which in turn increases the possibility of involuntary muscle firing.

Without going into too much detail, as my specialty is not medicine, we can relatively simply understand that what is at stake is therefore not pure dehydration, but something much more akin to compartment syndrome of the limbs. Again, very simply, this means that the muscles, when exercised, fill with blood and when the pressure inside the muscle compartment is too high, the swelling created leaves the blood with no way to escape, creating pain and blocking movement. In extreme cases, arm pump can lead to emergency surgery due to the risk of losing the limb in question, due to the "tourniquet" effect created by high internal pressure in the compartment - or body area - and the subsequent lack of blood flow to the extremities.

There are solutions to this problem ranging from stretching to preventive surgeries. However, knowing that excess effort in the forearms appears because we are supporting our body weight with our arms and not our legs, the tree analogy works perfectly once again, as it's impossible to support the weight of a tree by its branches without anything breaking.

FROM THE FEET TO THE KNEES

Understanding that fully planting our feet on the pegs creates problems and that bending the knees to circumvent them isn't an improvement, it becomes important to analyze the other benefits of choosing to position the foot in the toe-start area.

Armed with that mindset, we know that what we want is to absorb the irregularities of the terrain with minimal impact on our body, aiming to promote our longevity, safety, and comfort, while minimally influencing the natural workings of the bike. We also know that by moving our feet back we create a pivot point at the heel, and that this pivot point will solve our problems with impacts and imbalance.

However, even though this body adjustment brings a long list of benefits, it's not entirely free of user errors.

Image 6.11 – Good positioning of the foot and heel on the footpeg.

At the top of this error list is the tendency of many riders to try to keep the heel up and not let it drop, a decision - many times not conscious - that not only results in enormous physical strain on the calves and thighs but also inhibits the sought-after benefits. As this position with the heel dropped relative to the footpeg results in a hyperextension of some ligaments and muscles, it creates great discomfort for less flexible users, with some people reporting pain from the heel all the way to the buttocks, making this one of the reasons why many fall into the described mistake, even when they have the foot well positioned on the footpeg.

Nonetheless, as with everything, with continuous use of this position or, in other words, with training, this discomfort tends to disappear over time and is replaced by extreme comfort resulting from the benefits of adopting this approach. However, during this transition period between discomfort and comfort, it is possible to minimize physical wear by making a greater rotation between the various possible body positions - which we will outline further ahead - and by improving our flexibility, using sequences of Pilates exercises, like those found in Foundation Training. Still, less flexible riders and those without the time for this kind of physical homework will undoubtedly find a longer adaptation period to this technique.

With that said, it becomes important to explain how - when we can comfortably let the heel drop the body's natural spring is maximized, creating a new safety automatism.

By biomechanical rules, when we let our heel drop, our knees will be immediately pulled back from the perpendicular line made between the footpegs and the ground, becoming straight. This point is crucial, as positioning the knees forward of this line will increase the likelihood of them bending in an impact or distraction, resulting in the pendulum body problems described earlier.

At this stage, it is important to clarify that straight knees - or minimally flexed - does not mean knees locked in their most retracted position. If you stand up right now, you'll notice that although your knees are straight, you can still force them further back, taking them to the locking position, which, for safety reasons, we should avoid when on the bike, after all, something locked or braced does not have any give, and as such, is more prone to breaking under load.

Thus, although incredibly close, we find two distinct knee positions, and it is important, once again, to note that what we seek here is to keep the knees relaxedly straight and never locked. In this way, and as a bonus for keeping our knees straight and behind the footpeg line, we will immediately elongate the body on the bike, distributing our weight more effectively and getting into a position that more easily allows us to ride with an unlocked hip, a huge benefit for those who want to maximize their safety and control.

Image 6.12 – The heel aligning the body.

With this in mind, to ensure a good base position of the knees, we can and should use the heels as an automatism to align the whole body from our root.

HIP AND TORSO

There are many tricks and tips for improving off-road riding, but perhaps none is more important than the concept of body separation, to which we'll later add the sub-concept of initiation. This concept is based on the understanding that any bike can and should be seen as two separate entities, not as a single unit.

This approach to a bike comes from the fact that it's utterly impossible to maneuver the rear wheel, the most retracted and lowest point of the bike, with the handlebars, the most advanced and elevated point of the motorcycle.

Image 6.13 – Separating the bike into two through the steering column.

This is the case because, regardless of how much the handlebars move, we can never precisely maneuver more than the front wheel, with everything behind the steering column remaining basically unchanged, or moving in a towing scheme. This simplistic view of a movement easy for all of us to replicate has drastic implications on how we ride, as it requires a whole different approach from the rider.

That being said, what we want to avoid is maneuvering the bike by the handlebars alone, opting to lose control of the rear and leaving it in a train-like scheme, where the front goes and the rear follows. However, this approach - or any of its variants - is chosen by many riders new to the world of two wheels. Fortunately, this tendency usually disappears relatively quickly, being replaced by the habit of steering with the footpegs, a largely more correct technique, but still incomplete, as the majority of explanations of this technique tend to omit the concept of body separation, an important step to explain and understand.

So, if we accept that it's impossible to maneuver the rear wheel with the handlebars, we must also accept that it's impossible to control the front wheel only with the footpegs.

By pressing the right footpeg, for example, the bike will fall to the right, and to a certain extent, the front will automatically align itself, making it relatively easy to execute turns without handlebar inputs. This is a real phenomenon but a misleading one, and proof that "possible" is not synonymous with "well executed" or "complete."

Let's assume a wide slalom exercise marked with cones. As explained, it's certainly possible to turn without using the handlebars, using only the bike's balance given by an input to the footpegs. However, with this approach, we are using an "inverted train," where we give an input to the rear of the bike and letting the front follow. Now, if we shorten the distance between the cones in our imaginary slalom, it quickly becomes impossible to continue the exercise using the same technique. This happens because, to link tight turns, we will need to introduce handlebar inputs to increase and control the speed of transition between turns.

With this notion, we then arrive at the basic principle of body separation, where we are using our hands to maneuver the front wheel and our feet to control the rear wheel. However, understanding the necessity of using both feet and hands to control the bike is not enough, so it's important to expand the concept, as unfortunately, this is the point where many riders stop their analysis. I say unfortunately because, despite the mixed use of the handlebar and footpegs to steer being a correct and somewhat natural technique, it's usually applied in its most basic and, therefore, incomplete form.

Incomplete because it's not taking into account the natural biomechanics of the human body, resulting in common errors that are relatively easy to identify.

If you've ever seen someone get a jolt in the front wheel that resulted in the rear jumping off its trajectory, then you've seen one of these errors happening. Likewise, if you've seen someone hit the rear wheel and lose the front as a consequence, then you've also seen another one of these errors occur.

These issues arise because, despite the method of controlling the bike existing separately, the separation of the body is not being taken into account. This results in any input made with the hands being transmitted to the feet and vice versa. This happens because our body is acting as a conduit - exactly what we want to avoid. Armed with this simple analysis, we are forced to discuss and apply body separation, something easily achieved by the correct positioning of the hip, in a movement that works when riding both standing and sitting.

By sticking our butt out, or more eloquently, by rotating our hips outward, we can mitigate many inputs that previously traveled freely through the body. This, to a certain extent, allows the body to function as two separate entities, as this hip rotation that produces body separation offers immediate benefits and once again becomes a key point in developing yet another automatism.

Firstly, with this hip rotation, we automatically engage our abdominal and back muscles, straightening our backs and putting them in an active position, instead of having them tend to arch in a curled and defensive posture.

Image 6.14 – Difference between rotated hips and arched or straight backs.

While there's debatable validity to the subconscious fears that force us to curl our bodies into an almost fetal position while riding, there's no doubt that hip rotation allows us to achieve stability in our torso with minimal mental effort. This doesn't mean we shouldn't force even more body stability at certain times, like in certain turns, for example. It just means that we now have an excellent base position for the torso that we can easily achieve and maintain with a single movement.

This use of natural biomechanics follows the same theory as that of the feet, analyzed previously. However, it's important to note that it is certainly possible to keep our backs straight without rotating our hips. But as this is a more mentally draining movement and difficult to maintain in the long run, the tendency to lose this position is greater, even in high-level athletes.

Similarly, simply keeping our back straight, while promoting stability, does not promote separation, a key point to retain, as the position of our backs does not attenuate the communication between the upper and lower parts of our body in the same way. Another benefit of hip rotation is that, with a more intuitively and hip-based stable torso, it becomes easier to apply various techniques, whether in turns, jumps, or even speed variations, as we can attenuate the propagation of errors made by any extremity of the body.

This means if we can effectively create body separation, technical errors made with the feet are much less likely to destabilize the torso and vice versa. This allows us to drastically increase our safety envelope with a single body adjustment, or in other words, an error in the front or rear wheel does not influence the other wheel so directly.

By assimilating this clear benefit of safety and control, it becomes easy to understand the main problem of riding without body separation, as when the torso communicates with the legs, we are effectively connecting the front wheel to the rear wheel.

Let's imagine a turn where, due to poor foot positioning on the bike, something touched and pulled our foot off the peg. The foot pulls the leg, which in turn drags the hip, causing the shoulder and elbow to drop, and this new body posture makes the front wheel turn inward into the turn, often leading to a fall.

Riders with good body separation can – to a certain extent - more easily progress on the terrain without feeling significant body changes from these types of impacts, as the mistake tends to end at the hip and not progress as violently through the body. However, while body separation is an incredible resource, it can and should be enhanced with a simpler but equally important concept: initiation of movement.

A great way to understand initiation and body separation is through a simple practical exercise involving accelerations and braking in a straight line, something extremely familiar to riders of any level.

If you sit relaxedly on the bike, start riding, and once ready, accelerate consistently for a few yards, maintain the speed for a few seconds, and then brake back to your initial speed, you'll be greeted with a basic concept of physics: every action has an opposite reaction. In other words, you'll quickly feel that when accelerating, your body tends to be thrown backward, and conversely, when braking, projected forward.

Now it's time to introduce initiation, a concept that can be defined as preparing the body for what is about to happen. In other words, now with the hips rotated, if we're going to accelerate, we should lean our body forward before rolling the throttle, countering the movement that the body naturally will want to have, meaning, be pulled back. Conversely, before applying brakes, we should lean the body back before it is projected forward. The extent of our lean however depends entirely on how much we are accelerating or braking, making our initiation a constantly variable measure, not a predefined one. This point reminds us once again that riding should be a dance between us and the bike in a constant game of feedbacks from both sides.

So, repeating the exercise again, now with hips rotated and initiation, you'll immediately feel a difference in your stability, as well as the bike's.

Image 6.15 – Exercise with acceleration, speed control, and braking.

Another benefit of initiation with separation is that by having the torso straighter than we would in a naturally more fetal position, we're preventing ourselves from potentially going into apnea. In Chapter IV, when we discussed breathing control, we focused only on the mental aspect of it, but body posture is equally relevant.

When we're curled up with our backs, we've necessarily fallen over our chest, pressing it, making our breathing automatically more complicated and, as such, shallower and less deep, leading to a more rapid onset of apnea, marking it the perfect starting point of the fight or flight process we discussed earlier.

With all this in mind, I can't close this topic without mentioning that, while these types of techniques should be used as much as possible at the competitive level, for dual-sport, ADV, or leisurely enduro rides with friends, we can take another approach.

Due to the difference in rhythms and types of terrain typically encountered in leisure riding, it's relatively easy to find ourselves in terrains where for long periods of time we are in total control of the bike and the situation. Thus, because body separation is a quick technique to apply, whether seated or standing, using it only when necessary allows us to save energy, which is crucial when spending many hours straight on the bike, a clear example of how small changes in attitude and technique have significant impacts on our longevity and safety while riding.

ARMS

You've probably heard that to ride off-road safely - whether on an MX track or a leisurely Sunday ride - you need to keep your elbows up. This advice can be boiled down to the need of maintaining balance and stability. However, that explanation is as vague as it is helpful, demonstrating why clichés fall short in explaining this or any other technical movement. So, let's tackle this from a different angle and understand why poor elbow positioning disrupts our balance and stability to start with, a mental exercise that will much more help us grasp why applying good technique is crucial from the get-go.

To achieve our goal, we need to examine our body position in various riding stages, like on straightaways, in turns, and across different terrains. After all, we're discussing a highly dynamic sport, so opting for a single body position for all situations is a widespread mistake that often leads to issues, nevertheless, some fundamental concepts remain constant across all possible positions. Among these is the need to let the front do its work, avoiding micro-adjustments to align a direction that will self-correct - a task made much more difficult with poor elbow positioning.

The commonly found error of dropped elbow position, seen in riders from hobbyists to professionals, essentially locks our arms against our torso. While the consequences of this position are far-reaching, the problem itself is just that. Therefore, it's crucial to analyze the consequences rather than the problem itself, as the solution - lifting the elbows - is, as we all understand, quite straightforward.

Let's think of a T-Rex, the dinosaur we're all familiar with, but in this context, don't focus on its size or appetite, but rather on its tiny arms.

Image 6.16 – Poorly positioned elbows lead to a T-Rex-like posture, both laterally and frontally.

Biomechanically, it's impossible for a T-Rex to move its arms side-to-side without shifting its body position. Similarly, when we let our elbows drop and stick to our torso, we're essentially adopting the T-Rex stance. I've heard riders say they prefer this position because it lets them grip the front of the bike. However, not only is this counterproductive in terms of allowing the front to self-align, but the issues don't stop there. This becomes clear if we analyze the body position for a tight technical turn, for example.

The technique dictates that we should rotate our torso in the direction of the front wheel, maintaining stability and a certain parallelism between the torso, arms, and handlebars. With this definition in mind, imagine you're on a straightaway and receive a jolt to the front wheel that jerks the handlebars side-to-side. If your elbows are dropped and pressed against your torso, your torso will be pulled in the direction the wheel turned, causing you to lose bodily stability and your riding position.

This happens because the wheel's movement pulls the handlebars, which then pull the arms. Since the arms are locked to the torso, the torso is forced to rotate with the wheel, placing the body in an incorrect position for a tight technical turn, even though we're still trying to move straight ahead.

Thus, instead of allowing the front wheel to self-align through its trail, our body position signals to the bike that we want to turn, when in reality, we aim to go straight. This puts us in a situation that will likely result in a fall.

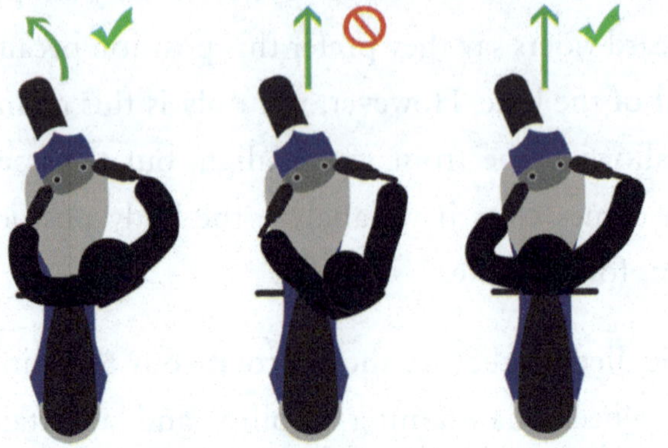

Image 6.17 – Difference between technical turn, impossibility of moving forward in T-Rex position with variation in front wheel direction, and forward movement with correct elbow position and variation in front wheel direction.

This type of fall becomes easily avoidable by simply raising our elbows, which separates our arms from the body, allowing them to move with the handlebars without disturbing our torso's position and the information we're giving to the bike.

On the other hand, keeping the idea of movement initiation in mind, combined with the T-Rex position, if we imagine instead of a side-to-side deflection of the front wheel, we have an impact forcing the suspension to compress, it's easy to see that the arms, being tied to the body, will force our torso backward with varying degrees of intensity. Depending on the rest of the body position, being thrown backward, the tendency for a good percentage of riders will be to grip the handlebars for dear life. Given the position of the torso and arms, this promotes a whiskey throttle, which will force the body to fall even further back, leading to a fall.

Alternatively, and again depending on the rest of the body position, many riders choose to momentarily release the throttle. This means they will be pulled back by the impact, will be projected forward as they reduce speed, and, upon returning to the throttle, will be thrown back again, creating the already discussed ragdoll effect. Therefore, by raising our elbows and adopting an arm flexion position relative to the handlebars, we can absorb impacts suffered by the front wheel through a shock absorber created by the elbows. This allows the wheel to move up and down in the suspension travel, with minimal impact on the torso.

In this way, we can let the bike work underneath us without having to change our body position, thereby maintaining the control, consistency, and safety we seek in our riding.

HANDS

A typical technical mistake seen in riders ranging from professionals to hobbyists is the lack of application of an extremely simple technique known as "re-grip." In leisure riding, this flaw is particularly visible when transitioning from a turn to the next straightaway.

Those who ride on highways or any long stretches where the throttle is significantly engaged for extended periods of time are familiar with the concept of releasing and re-gripping the throttle in a new position. This move allows the rider to maintain the desired acceleration level while promoting a more comfortable hand position, reducing wrist torsion and tension.

This technical gesture, which is literally known as re-gripping, helps to remove the arm from a position where the wrist is bent, which forces the elbow downward, which in turn, reduces the stability of the torso. Additionally, not maintaining our forearms and hands in a straight(ish) alignment with the grip, by having instead an accentuated bend at the wrist, facilitates the emergence of injuries, promoted by even the smallest impacts.

Image 6.18 – Poor wrist positioning on the handlebars.

Applied to the exit of a turn, for example, this mistake - by altering the elbow and torso position - increases the risk of encountering all previously described body instability issues, once again reminding us that our body must be seen as unit, and not as separated parts, even when analyzed as such.

To enhance our safety and control, we should then raise our elbow, which in this situation is impossible without readjusting the hand's position on the throttle.

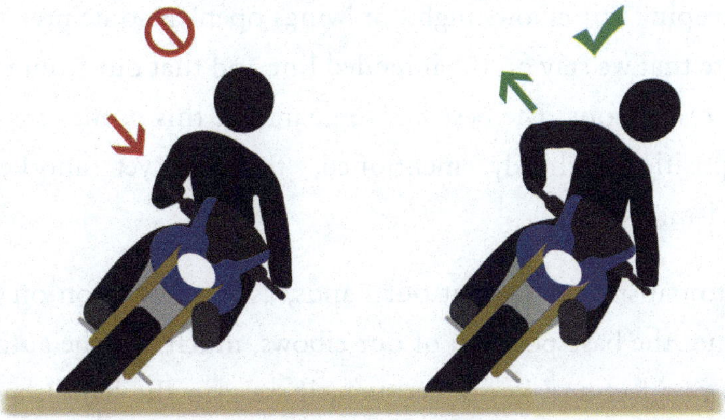

Image 6.19 – Poor outside shoulder positioning in a turn.

However, it is important to note that re-griping can take it's time as it is a maneuver that requires preparation, making it impossible to execute instantly on a whim. Therefore, the way it's applied varies with some riders choosing to perform it before or during turns, with others choosing not to do it at all in certain specific situations. Nevertheless, not performing it doesn't necessarily mean losing complete body posture or elevating the elbows as much as possible. It just signifies that by choosing not to use this technique, these riders should be fully aware of the risk they are taking, which drastically changes their approach to the movement.

Therefore, we must bear in mind that being aware of this technical movement gives us choices, something crucial to eliminate persistent errors and falls. At this stage, it becomes easy to understand why keeping the elbows high allows the front wheel to work laterally and along the suspension travel, and also why it promotes the stability of the torso - effects impossible to achieve while adopting the T-Rex position.

So, by keeping our elbows high - or 'wings open' as some prefer to say - we ensure that we stay on the intended line and that our front can work without restrictions. The best way to guarantee this position is, like the other positions already mentioned, through yet another body automatism.

This automatism is all about our hands, as their position on the bars determines the base position of our elbows, much like the automatism between our feet and knees. As such, if we grip the handlebar with a "full hand," as we tend to do due to the handlebar's sweep, or the angle at which the handlebar turns toward the seat, we'll end up with our elbows resting too closely to our body, which is counterproductive.

In contrast, if we rotate our hands out relative to the handlebar, making a sort of diamond shape with our hands, we'll grip the handlebar at an angle that allows the rotation of the radius and ulna in our forearm. This bone rotation along our elbow allows it to be positioned higher and further away from the torso in a more relaxed and automatic manner.

Image 6.20 – How different handlebar grips influence elbow position.

As an analogy, for those who have tried doing diamond push-ups, the hand position is similar. This rotation of the hand, like the other automatisms we've discussed in the past, doesn't prevent the need to lift the elbows even higher in specific situations. This position simply provides a guarantee of a minimum safety position without having to spend mental capacity, which in many cases, is enough to promote more consistent and safer riding.

Just as this slight rotation of the hand on the handlebar offers the benefits already mentioned, it also provides greater sensitivity in terms of fine throttle control. Our wrist, as anatomically impressive as it is, isn't known for its extreme sensitivity - a quality abundant in our fingers and hand.

Thus, by avoiding grabbing the throttle "full-handed," which would mean controlling our speed choice with our wrist, we find ourselves gripping the throttle in a position much more similar to how we would turn a round doorknob or an ice-cream. This position greatly increases our sensitivity on the throttle, which in turn, significantly improves our control of the bike.

However, it's always important to remember that the bike's throttle isn't the means used to transfer power to the wheel; the throttle is merely a suggestion of power we're giving to the engine.

This concept might seem complex and even wrong, but if we consider that we can get our throttle wide-open and, if we pull the clutch, the bike won't move, then we must accept that the throttle tells the engine more or less what RPM we want it to work at, with the clutch position dictating exactly how much power is transferred from the engine to the wheel.

While understanding this concept is crucial for fine-tuning speed in various maneuvers, its execution is nevertheless improved when the throttle position is more precise, something impossible to achieve if operated based on a body part deficient in sensitivity and fine control.

LATERAL AND LONGITUDINAL MOVEMENTS

Off-road riding is far from static; it's an incredibly dynamic activity, which is why understanding how we can and should move, now that we have a defined base position, is vital.

To fully grasp our possible movements, we must distinguish between two distinct types: movements along the bike's line, or longitudinal, and movements across the bike's line, or lateral, keeping in mind that diagonal movements, formed by combining the previous movements, are what we tend to use most. To understand these movements, and remembering the need to be as precise and to the fraction of an inch as the bike in how we move, we must define clear rotation points along our body.

You've probably heard about the need to shift your body back or forward to maximize traction and distribute weight. While this requires a deeper understanding of what's happening with our wheels - which we covered in Chapter III - it also requires knowing exactly how to move our weight while riding in order to avoid destabilizing the bike.

Thus, with our base standing position in mind, and our heels as our first point of rotation, we can define the knees as the second and the hip as the third, remembering that the hip is one of the most crucial points in controlling the bike. Not forgetting any of the hip related points debated before, we must keep present the notion that the hip, being the center of our body and the point of body separation, can significantly change the ride/bike center of gravity by moving just a few inches in any direction. This big shift with minimal movement is due to the immense weight of the rider that is being moved relative to the bike, thus creating a high leverage arm relative to the center of the vehicle, directly and clearly impacting its stability.

By using these three points, we can easily alter the maximum height of our vertical body pendulum, as well as vary the positioning of our "upper mass" vertically and longitudinally.

For many, this way of moving, based on a relatively fixed point at the knees, seems limited because it gives the impression that we're not maximizing movement. However, I must disagree with this view, as this simplistic approach doesn't take into account a maxim previously mentioned: our task on the bike is to give it the best possible working conditions and, as such, move without destabilizing it.

If we keep in mind that, on average, about 60% of our weight is situated from the hip upwards, and that this weight, especially when standing, is elevated, it creates a considerable leverage arm relative to the bike's center of gravity. Therefore, any movement we make is never moving little weight, quite the opposite, forcing us to be precise in our movements, moving just enough to give the bike what it's asking for without the exaggeration of a large mass transfer that can - and usually will - destabilize the vehicle.

This means that, to increase our precision in mass placement, we should move as much as possible without changing our knee position, keeping the root and beginning of the trunk of our body tree as stable as possible. As such, to put more weight to the back of the bike, we should shift our butt back, using our knees and hips as points of rotation. To transfer weight forward, we must move our torso down and/or forward, rotating it around our hip.

Image 6.21 – Body rotation points with weight movement back and forth.

Using this type of movement allows us to not only work more easily the initiation needed for technical movements but also to stabilize the bike, allowing us to make the most of the suspension and geometry of the vehicle.

Obviously, the dynamism of these movements is defined not by the maximum amplitude we can achieve, but by the needs determined by the terrain and throttle position. Once again, we must keep the idea of a dance in mind, forcing ourselves to give the bike only what it really needs and asks for, and not what we assume it needs before having its feedback.

Thus, being aware of how to move and understanding Kamm circles, we can safely, with minimal physical input and maximum technical impact, increase and reduce our effective traction circles by precisely placing our body mass longitudinally.

Image 6.22 – Weight movement and impact on Kamm circles for each wheel.

Regarding lateral movements, these are exactly as the name implies, movements of the body lateral to the longitudinal line of the bike, which must be divided into two basic and, in a way, complementary movements.

Both necessary movements follow the same logic, which is the need to tilt the bike to one side, compensating for its fall by letting the body move to the opposite side, ensuring that the center of gravity of the bike/rider set remains central. Like other movements, how much we need to "come out" of the bike is determined by the inclination and speed of the maneuver, so it must be precisely adjusted in motion and not predefined.

Image 6.23 – Lateral weight movement to compensate for bike tilt at low to medium speed.

To achieve this body position, our intuition tells us to step on the footpeg opposite the direction of the turn, to compensate for the tilt of the bike, a concept that is fundamentally correct. However, its effectiveness has problems if we leave the explanation and application there. This is because, by telling our body to step on the footpeg, we will tend to use our whole body to do it, applying a body movement that goes from feet to shoulder, something that, by definition, will make us lose torso stability, leading us not only to exert more force than we probably need but also to overly destabilize the bike.

Therefore, it is important to create an automatism of movement that, in this case, is linked to a lateral hip movement. By pulling the hip to the side, we effectively divert our weight, forcing us to exert more force on the footpeg on the side where we moved the hip to, thus achieving our initial goal without the need to lose stability and control.

This does not mean that there are no exceptions regarding the amplitude and the violence of application of some of these movements.

After all, in the real world, distractions, technical or mechanical failures, and a whole array of possibilities affect how we must move. Nevertheless, it becomes important once again to remember that we should always follow a simple mental checklist regarding body movements to avoid exaggerations and improve consistency.

First, we should initiate each movement based on our central body position, then apply the best possible technique possible to what the bike is asking us - whether that is a linear or exaggerated application - and then, as soon as possible, return to our central position to repeat the process.

This simple checklist, combined with the simplification of movements explained and supported by a present state of mind, is the basis for consistent and, in turn, safe riding. This is true because off-road on two wheels is a short-memory sport where we must be focused on the present to prepare for the future, but never live in the future while surviving the present, or dwelling in the past. As an example, we shouldn't hang on that mistake or scare we just experience, nor should we stay focused on that difficulty that we may know will eventually show up ahead. We should ride in the present preparing the immediate future, because that is where our bike is and will shortly be passing at speed.

To this mental state, if we add fluid riding free of exaggerated and abrupt movements, and allow ourselves to dance with the bike, instead of trying to physically overpower and force it to do what we want, we find the recipe for the vast majority of any rider's needs, whether on Sunday rides or world championships. Once again, this does not mean that there are no exceptions, obviously, there are, but this method also foresees them.

This is because, whenever our procedure memory and checklist are not enough, when our process is consistent, we will be able to understand exactly what failed - and how to adjust it - by understanding how much we are deviating from both the bike's comfort zone as well as ours. This understanding will be crucial not only as self-preservation but also as a method of constant and continuous learning.

As a bonus of this application of concepts, our procedure memory will become increasingly robust and, as such, we will find ourselves effortlessly expanding our comfort zone, thereby increasing both our speed and the difficulty of the terrains where we can ride safely.

THE LIMITS OF THE BIKE AND THE BODY

At this point, it's entirely reasonable to wonder about the limits of longitudinal movement, as I haven't strictly defined them. I've mentioned that we shouldn't move our knees too much, pointed out exceptions, and emphasized responding to the bike's demands. As such, these limits remained undefined by design because they are largely dictated by two primary and highly variable characteristics: the type of bike we're using and the physical attributes of the rider operating it.

An adventure bike, with its typically wide and split seat, will naturally restrict how far back we can shift our weight when compared to an enduro bike, for instance. Likewise, on the same bike, a taller rider will be able to move further back than a shorter one - and it's not just about leg length - as our arms also dictate the limits of our movement, considering we should never lock or straighten our elbows completely to maintain control of the front wheel.

Keeping all this in mind, we quickly realize that these imposed limits for moving backward apply equally when shifting our weight forward. On an adventure or rally raid bike, which features an instrument tower and a windshield, we'll reach the forward movement's limit sooner than on an enduro or motocross bike, which has no obstruction to the forward movement of the head and torso.

Thus, with these and other possible examples in mind, it becomes crucial to understand that the concepts explained here should be carefully applied to each individual's reality. This reality is influenced by your size, bike, physical ability, and riding objectives.

BASIC DYNAMIC MOVEMENTS

We've already established a base body position and introduced the concepts of initiation and separation. However, we've considered these aspects in a relatively static manner and, even when addressing movement, haven't fully taken into account the terrain variations we tend to encounter in the field. To have a simplified and easily replicable checklist for all scenarios, we must then discuss uphill's, downhill's, turns, and braking. But first, let's define a crucial concept in the base of all of these: momentum.

Regardless of the type of bike we are riding, maintaining a consistent speed - or introducing gentle variations - in any terrain or situation is what allows us to let the bike work at its best. By avoiding regular use of harsh braking and aggressive acceleration as a riding method - movements that inherently create instability in the bike and require more physical effort to compensate - we promote a fluid movement of the bike/rider unit, in what many define as a state of flow. This state usually refers to how a rider is physically and mentally handling certain situations, or in other words, when the rider seems relaxed yet in total control during a specific section of the course, regardless of its difficulty. However, we often forget that this state is only attainable when the bike is also performing at its best, a state of the vehicle for which we bear 100% responsibility.

By understanding how to use the bike's momentum to our advantage through a basic checklist, we become more consistent, which in turn facilitates the bike's operation, which facilitates our entry into a potential state of flow. Considering that momentum is calculated as mass times velocity, and knowing that our moving mass remains constant (considering the consumption of gasoline irrelevant for this explanation), the only thing that varies our momentum is our speed.

Knowing that braking adds additional load to the front suspension and accelerating does the same to the rear, we can compensate for both to a certain extent with our choice of body position. Thus, we can assert that each of us has a comfort speed, a riding speed that requires minimal physical wear and promotes bike stability.

This comfort speed is where we can save the most energy by riding in a way that requires little physical effort. So, we can broadly define that in competitive terms, we want to conserve momentum to lose as little time as possible, but in leisure riding - where we fit in - we want to conserve momentum to save as much energy as possible.

To conserve momentum throughout a track, we should aim to take our turns at roughly the same speed as our straights. This means that in addition to conserving energy by not constantly making major speed and body position changes, we also tend to be faster overall, as well as experience fewer scares and problems along the way.

However, applying this concept tends to mean adjusting our cruising speed, typically by reducing straight-line speed or, at least, the speed of entering and exiting turns. A good method to achieve this is to reduce our maximum comfort speed by about 15%, which will promote a more consistent overall speed, offering natural and sustained growth in our general speed.

Similarly, we should avoid abrupt inputs on the bike and should always be smooth and exponential in applying our controls, whether brakes, clutch, or throttle. This approach promotes a gentle initial 'tension' in the bike that supports a more aggressive application of controls in subsequent moments, greatly enhancing our precision in control application while reducing bike instability.

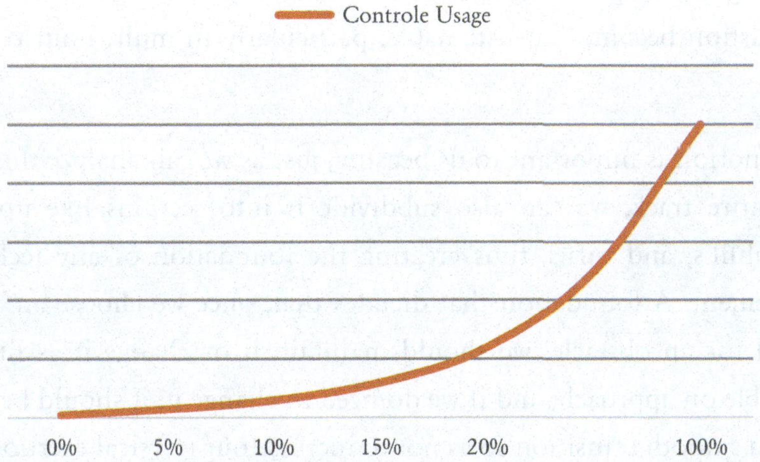

Graph 6.1 - Being smooth in the first 10 to 20% of command application enhances the exponential application of the commands for the rest of their range.

In summary, we can say that less fluctuation between maximum and minimum speed allows for a more precise application of procedures. This, in turn, places us in a position where we can execute the necessary movements in a more fluid and mechanical manner, and thus, faster. This approach, without seeming like it at a first glance, does increases our overall speed in time without us ever having specifically sought to do so, as when we do so, we do it incrementally everywhere, and not just on straightaways.

Conversely, riders who seek higher speeds often disregard these basic rules and tend to over-apply the throttle beyond their real riding capabilities. This approach frequently leads to aggressive braking for turns or "overcooked" turns, followed by exits with little to no control and consistency. This method of riding comes at the cost of enormous physical wear, a poor sense of safety limits, potentially more scares than needed, which in turn promote mental and physical wear, and overall, lower general speed, especially over longer stretches where physical exhaustion becomes unsustainable, particularly in multi hour or days rides.

This notion is important to us because, just as we can analyze this over an entire track, we can also subdivide it into sections like uphill's, downhill's, and turns, thus creating the foundation of any technical movement. A foundation that dictates that, once we choose an attack speed for an obstacle, we should maintain it or change it as little as possible on approach, and if we do need to change it, it should be done with a smooth transition so as not to increase our physical exertion and to maintain the bike's stability during the movement.

UPHILLS

We've repeatedly talked about the concept of dancing with the bike and, in a way, waiting for the bike to tell us what it needs before we assume we know what's necessary to do. This is a factual reality, yet it doesn't mean we should completely eliminate our decision-making process and discard our cognitive ability to analyze problems while riding - especially as our experience grows.

Knowing that on an uphill, the tendency in analysis - somewhat exaggerated - for many will be for the front wheel to want to lift off the ground and the bike to pull a wheelie, it's easy to assume we should put our weight forward as soon as we tackle the ascent. But doing so would bypass our analytical capacity and the bike's requests.

This is because, if we're on a flat surface and have an uphill ahead, what's going to happen is that the front wheel will start the ascent before the rear wheel, inevitably causing the front of the bike to rise towards our body, with varying degrees of aggression, depending on the initial slope's steepness.

If on enduro MX or SX bikes this is usually not a major problem due to the lack of obstructions between the handlebars and the fender - and the way the tracks are designed - the same cannot be said for adventure bikes or those prepared for rally raid, for example.

Image 6.24 – Example of the front wheel rising towards the rider due to the change in the bike's attitude when entering an uphill.

Here, if we prematurely position our body forward as a method of forcing the front wheel down, the bike, upon starting the ascent, will tend to want to collide with our body, usually hitting the helmet or neck, putting us in a precarious situation.

Likewise, not all ascents are the same, so, even with both tires on the uphill itself, there might indeed be a need to increase the load on the front tire, in the same manner than in other cases, we will need to do it to the rear tire. Thus, if we've prematurely decided to be in a forward body position, if needing to quickly shift back, even momentarily, the movement will result in a rapid mass transfer that will destabilize the bike, putting us in a situation we know to avoid.

So, my advice in terms of body position for tackling uphill's is to keep the body in a central position, with a slight tendency forward, ensuring that our elbows remain high. This way, the arms will be able to absorb the front of the bike rising at the start of the uphill and, if we need to put weight on the front or back wheel, that mass transfer will be smoother and quicker, promoting a sequence of precise and fluid movements.

Regarding the momentum we mentioned earlier, this is a concept we should always keep in mind. This is because we should choose the speed of attack before reaching the base of the slope, allowing us to accelerate smoothly and consistently throughout the uphill. This acceleration should not be done in bursts or by fully twisting the throttle. The idea is to linearly compensate for the rotations lost by the engine during the ascent, with the possibility of increasing speed if necessary, reducing the probability of loss of traction in the process.

On one hand, not having enough speed to climb will necessarily force a failed maneuver. On the other hand, any abrupt variation in acceleration to compensate or increase speed during the ascent will disturb the bike and cause the rear wheel to want to skid, especially in looser terrains, so smoothness and fine control of the throttle and clutch are as fundamental here as in any other dynamic situation.

With the idea of using the clutch as a method of fine control of how much power reaches the rear wheel, as we approach the top of the hill, we should maintain the throttle - within certain limits - and cut the bike's power by pulling the clutch, not cutting the throttle, ensuring that our arrival at the top is done almost solely by maintaining the existing momentum. This maneuver is purely about safety, as it safeguards a variety of possible scenarios.

By definition, in adventure settings, we tend not to know what we will encounter ahead, so by reaching the top of an uphill, it's common not to see the next few yards, which can have holes, ruts, logs, animals, or even a simple terrain variation that promotes a jump or a fall.

Thus, reducing our speed near the top will allow us time to analyze the situation and choose the best attack plan. However, if we slow down by cutting the throttle, we will necessarily reduce the engine's RPM's, making it difficult for the bike to react if we immediately need available power once again. Conversely, by cutting power to the wheel by pulling the clutch, but keeping the throttle in a runningish position, we can effectively reduce speed without a break in engine rotation, allowing us to quickly reintroduce speed to the bike if needed without the risk of engine stalling or forcing a potential loss of traction in the rear wheel.

DOWNHILLS

When tackling a downhill, it's a bit of a reverse approach compared to an uphill: we should aim to stay centered, but with a tendency to lean back. A quick look at various downhills shows that on many of them, we can stay centered without any issues, like losing traction or braking capability. However, some downhills present steps, whether at the beginning or somewhere along the way.

On a step, the front wheel will essentially disappear beneath us, dragging the rest of the bike with it, meaning that there is a case to be made for overcoming some steps with a jump, but that's a more advanced maneuver and beyond the scope of this book. So, assuming the front wheel will navigate the step instead of leaping over it, if we've assumed a more rearward body position, we might find ourselves with limited arm reach to handle the bike's lengthening over the step. This increases our likelihood of being yanked from our position by the handlebars, to the point of potentially being thrown over them. This is just as real a possibility for a step as it is for a hole or any other depression we might encounter on the downhill surface.

This being, and in a similar fashion to what we did on uphills, maintaining a central position - but now with a tendency to lean back - allows us to navigate any terrain quirks with control and stability through good body position.

Unlike uphills however, downhills should be started at a reduced speed, especially when we have limited visibility ahead or are uncertain about the difficulty of what we're seeing. Remembering that everything we do on the bike should be inch-precise, by keeping the brake lever and pedal preloaded - or in other words, pressed just enough to eliminate the normal working slack - we assume a position where a fraction of an inch more of brake application will immediately stop the bike, and a fraction less allows us to safely increase our speed.

This approach often results in a particular squealing of the brakes, which can - and should - serve as an indicator that we're in the desired brake control zone. Similarly, this method of controlling commands to eliminate slack can be used in any risky situation to increase our safety envelope.

Although we should always ride with one or two fingers on both the clutch and front brake, in a situation requiring quick reaction, our brain will request a percentage of braking that it calculated as being an X% measure. If our fingers aren't on the levers, or we haven't removed the levers slack, nothing will happen immediately when we start to pull the lever, even though it's already being actuated. This is a reality down to a fraction of a second, but it's enough for our brain to assume an error in the percentage it calculated, prompting it to ask for more, resulting in excessive braking - the same can be said for fine clutch adjustments.

This issue doesn't arise when we have time for analysis and adjustment, but it's an undeniable reality when dealing with quick, automatic reactions. That's why we should protect ourselves preventively and preload the bike's control instruments whenever we're entering more delicate situations.

Having reached a precise point in using the controls, we must consider a possible readjustment of the central body position, as extending our right foot forward to activate the brake pedal can lead to a tendency to rotate the hip to the left, giving this direction indication to the bike.

While this tendency exists, it's not mandatory, as it's perfectly feasible to execute the movement of pulling the brake foot back and forth with minimal movement of the right knee and thus, with an imperceptible or even nonexistent hip movement. However, this movement depends on the position and working range of the brake pedal.

A pedal set much below the footpeg line, or with a large working range, will inevitably force the right knee to advance excessively from its normal working position to allow the foot to activate the brake. For this reason, adjusting the bike to suit our own needs is absolutely crucial, as it is accepting the bike's adjustment limitations as a way to understand the limits of technical movements possible with that machine.

With the bike and body balanced and the brakes pre-activated with precision, we can begin our downhill descent, keeping in mind that we can always increase or decrease our speed as a result of our terrain reading. Braking, contrary to what many assume, should not only be done with the rear brake but with both, as the front brake remains the most potent braking method we have on the bike. However, engine brake should also be used to assist in the maneuver to a certain extent.

Although the notion of braking with the engine by varying the clutch position is simple and intuitive even for beginners, using the front brake in a downhill - or in off-road riding as a whole - tends to be a bit more abstract. Here, our understanding of the bike's traction and geometry, as well as basic physics, deepened in previous chapters, comes into play. Thus, knowing that front braking will compress the front suspension, which on a downhill increases the tendency of the rider to be thrown over the bike, the only thing we need to keep in mind is to ensure that our braking is progressive and not aggressive, as a way to gradually control the energy being placed in the front suspension. By maintaining pressure on the knees against the bike, we can counteract the bike's imbalance caused by braking, pulling our weight back through the hip. Once again, this good body position frees the arms, allowing the front wheel to read the terrain without our body being influenced by the handlebar position.

In this situation, the effectiveness of various ABS systems dedicated to off-road riding varies, with the most recent ones tending to have high levels of success. However, understanding how to play with the bike's mechanics will increase the knowledge and fine control of the commands for anyone, which is why I am a strong advocate that we should learn without electronic aids whenever possible. In this way, when we return to practicing with aids, we will have a much clearer understanding of how that particular bike's electronics works and how that limits or assists us as riders.

That said, if our fine clutch control serves as traction control, allowing us to precisely cut or give power to the rear wheel, our fine brake control and the sensitivity to momentarily release them when the wheel starts to lock will serve as a manual ABS. A clear example of this is locking the rear wheel on a downhill as a method of speed and position control.

Although this technique is chosen only for specific situations, when applied, this technical choice often leads to the rear changing trajectory, veering to one side. At this point, as riders riding with good body separation and endowed with all our analytical abilities, we must make a decision. Either we use our hip to prevent the rear wheel from passing by our side - in a hip movement opposite to the bike's intention - or, feeling that we need to realign the vehicle, slowly releasing the rear brake, allowing the wheel to roll again, regain traction, and self-realign, reaching the moment we can lock the wheel again, resuming the movement.

Of course, not all downhills or uphills require a technical application that blindly follows these patterns, but if we have these technical movements as a base, we can always choose not to apply them. If we choose not to have a base movement for each situation, our choices will not follow a replicable sequence, making our riding inconsistent and, as such, more prone to mistakes. We must remember that the professional athletes we idolize are not just riding bikes, they are operating vehicles in a systematically careful manner, which is why they can replicate laps within the same second, as well as execute a variety of maneuvers with high precision and consistency.

This discussion opens up the debate about whether, as leisure riders, we will lose some kind of fun or freedom by training to operate our bikes with a high level of systematic precision, but for me, the answer is clear. Having information and knowing how to operate at a high level does not obligate me to do so 100% of the time, leaving the door open for days of pure hooliganism or unmeasured clowning. However, even on those days, my training gives me a clear understanding of risks, which in itself increases my safety even when I choose not to apply what I know. On the other hand, I find immense pleasure and fun in trying to be better than I was the day before, but that, of course, is just my personal two cents on the subject.

EMERGENCY BRAKING

By this point, I have no doubt many of you are drawing parallels between our downhill posture and what's expected in terms of emergency braking, and if you are, you'd be right to do so. This is because, indeed, there are many converging points between these two technical maneuvers. Yet, they are different and, as such, should be approached distinctly, with emergency braking requiring some particular precautions.

Parallel to downhills, we have the need to pull our body back as a method to compensate for the energy transfer going into the front suspension, as well as the need to ensure that despite our weight being back, we don't do it to the extent of locking our arms, preventing us from absorbing impacts on the front wheel.

We also need to be precise and controlled with the brakes, always applying them exponentially. Thus, we should start their application in a precise but careful and measured manner until they show the first signs of action. At that point, as there's already tension in the front suspension, we can increase the force of application exponentially. This way, we promote greater stability of the bike as well as control on our part.

However, this is where the parallels end, and the specificities of emergency braking begin. This is because emergency braking is exactly what the name describes: a situation where, for some reason, we find ourselves needing to stop as quickly as possible, with potentially high risks if we don't do so.

This point becomes crucial, especially in training, as many, out of sheer ease, decide during the maneuver not to come to a complete stop, reaching the point of almost stopping and then immediately restarting motion to repeat the exercise as quickly as possible. Others tend to decide to release the brakes near the stopping point as a way to promote a smoother and less abrupt stop.

I understand how the precision of what I'm about to say may seem irrelevant in the face of two apparently "harmless" attitudes. However, this precision is vital to ensure the success of the maneuver when we need it most, i.e., in a real situation.

Therefore, it's vital that when we're training this technical movement, we do so until the bike is actually stopped, even if that means the vehicle falls at the end of the movement, or that we do fewer braking attempts per minute.

This is because every movement we're training is creating a replicable pattern in our brain, a procedural memory that we will follow in automatic mode when needed. This means that if we practice emergency braking to almost stopping, when we need to apply this knowledge in the field, we will almost not hit whatever it is that forced us to violently apply the brakes in the first place.

Another common mistake we must avoid is the fear of using the front brake, the brake that will effectively allow us to stop.

While the application of the rear brake is tied to stabilizing the bike, moving our weight to the rear isn't intended to mainly increase the traction of the rear tire and thus its braking capacity. Instead, it's about trying to prevent our weight from being violently projected forward when we violently activate the front brake, allowing us to stabilize the bike by a good distribution of load across the suspensions.

The rear suspension receives an increased load from our weight shift, and the front suspension's load increase comes from the energy transfer during braking. Keeping in mind everything we've discussed about bike stability and how weight and body attitude variations affect the Kamm circles of each tire, when we pull our weight back and lock the rear brake, we give the bike a "wheelie" posture. This is immediately compensated by braking applied on the front wheel, effectively stabilizing the vehicle during emergency braking.

Using engine braking is undoubtedly a great help in this maneuver's consistency. However, this step can be skipped in the initial training phase because braking without good clutch control - a completely different procedural memory - means the worst that can happen is the engine stalling, which will effectively lock the rear wheel, something we're already trying to do.

There are definite benefits to performing emergency braking without losing the ability to accelerate at any moment - impossible if the bike stalls - and even without locking the rear well all together, which is why in more advanced braking exercises clutch control to maintain engine rotation and downshifts are required. But, as mentioned, this is for more advanced exercises.

So, in a checklist format, we can see the steps of emergency braking as: move the weight back, apply full rear brake without releasing, pull the clutch (if possible, within your technical capabilities), followed by a tense and exponentially application of front brake, until the bike is stopped.

Textually this process may seem long, which doesn't make sense when we talk about an emergency maneuver, but don't be fooled. We're talking about fractions of a second in executing these steps, even for riders with limited hours of training.

We could create different sequences for bikes with electronic aids, but the specificity of each of these checklists for each specific bike would limit us as riders. By applying this sequence, however, we'll have the same success on any bike, regardless of the particularities of its braking systems.

Likewise, by having this procedure well-defined in the brain, we'll quickly gain processing time to add downshifts and clutch control, whether to keep the bike running or to decide to abort braking and accelerate through an escape route we find.

TURNS

Understanding the basics of how our tires' traction works and the dynamics of motorcycle operation, it becomes relatively simple to understand that, to a certain extent, we can turn off-road in the same way we do on road in recreational modes, that is, without moving our body all that much.

However, this procedure tends to only have a good safety margin in very specific situations, which is why it should not be the preferred method to adopt as a base.

Therefore, I always advise training turns in their most demanding regime, which will not only guarantee success even in the most complex situations but also allows us to establish maximum limits. This means that on the terrain, it will be easier for our brain to simplify a well-known movement, rather than making it more complex when we're in a tough spot.

Thus, training tight 180-degree turns, tight slaloms, as well as figure-eight direction changes, all at excruciatingly low speeds, is my recommendation for the most part, as this training model will require excellent clutch and throttle control, clear and precise front brake control, as well as an excellent body position.

The body position for these turns, considering we're talking about turns made at very low speed, does not require us to keep our heel dropped as we saw we should have in our base body position. First, because at the execution speed, we won't really have major impacts to absorb, but also because, by forcibly keeping the foot in that position, we would be limiting the work of the rest of our body for this movement.

This is because the turning movement should not be given by the handlebars or the footpegs, with these being a byproduct of the movement initiated by the hip, as we saw in lateral movements. Thus, the procedure to use and memorize involves little more than giving a knee nudge in the direction of the turn, and allowing the rest of the body to adjust to this movement without limits or constraints. If you ever did any kind of combat sport, or have skied, this knee "kick" rotation will feel extremely familiar.

If we imagine a turn to the left, what we want to do is rotate our hip in a knee-nudge movement of the right knee against the tank. This is basically the only movement we should have in mind, and one that works exactly the same whether we are seated or standing.

This is because, by rotating the hip to move the knee forward in the direction of the turn, we rotate the torso, which points the outer part of our body to where we want to go, which promotes the bike's lean that, due to the rotation of the outer part, allows the arms to follow the handlebar, automatically leading the body to step on the outside footpeg of the turn as an innate matter of body balance, supported by the leg that gave the knee nudge and initiated the rotation of the whole body to start with.

Now, if standing this movement is fluid and free of obstructions, sitting down requires an extra step, one that requires a clear acceptance that, as riders, we are an active part of riding and not just a passive passenger. Thus, to allow the rotation of the body through the sequence that was explained, we necessarily have to "lift off" our butt from the seat to allow the outside knee to move freely and the hip to rotate as needed. This movement on the seat can be taken to an extreme where the body is in a position where the buttock on the outside of the turn is off the seat, leaving us seated on the remaining butt cheek, or even already on the leg on the inside of the turn.

With all this in mind, it's crucial to remember that this automatism requires a good understanding of all the movements and body positions discussed so far, as well as a physical and mental relaxation that allows us to follow through the maneuver smoothly. But just as we need to learn our ABCs before we can read and write, by definition, a more complex maneuver like this always requires good learning and understanding of the basics.

That said, this movement allows us, at the limit, to achieve turn inclinations that biomechanically make it impossible to keep the inside foot on the peg during the turn, forcing us to maintain all our weight on the outside peg. However, this does not mean we lose control of the bike. It only means that, instead of being anchored by the knees and ankles, we might have to anchor ourselves to the seat by the outside knee and the calf of the inside leg.

Image 6.25 – Maximum turning position by shifting weight laterally to compensate for the bike's lean.

By managing to execute a movement as extreme as this, we reach the limit, creating our maximum procedural memory. This will then allow us to execute it promptly in the field with the same ease as we would a simple running turn. This is because now our brain will have a greater capacity to naturally realign us to compensate for any variation in the turn or attack speed.

However, it's important to emphasize that, in tight turns, the position of the foot may - and should - change on the peg to allow for body alignment. But in more flowing turns, if possible, the foot can perfectly remain in its base position, meaning resting by the balls of the feet - whether seated or standing - while still maintaining the idea of initiating the turn through the hip and knee, and not by directly loading the outside peg.

This is because directly forcing the peg will imply a dip of the shoulder on the same side, as a natural response to wanting to move as much weight as possible, something that will disturb the overall body position, creating problems of control, consistency, and safety. These problems are entirely avoidable if we opt for the movement of the hip and knee, which in turn will generate only the necessary load on the peg without disturbing the body's stability.

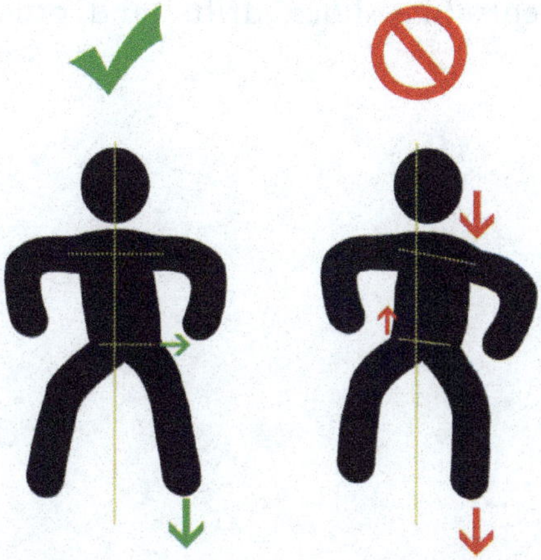

Image 6.26 – Differences in body stability alteration by different methods of increasing load on the foot.

Yet, adopting this automatism, like all the others talked about so far, does not mean that we are totally incapable of exaggerating this movement. It only means that if its base application is correct, its exaggeration will not directly imply the emergence of errors, with the exaggeration serving more as an ace up our sleeve that we can use when necessary.

When talking about turns, we must also remember the idea of counter steering, a movement that has as much applicability off-road as on road. One that can and should start to be mentally introduced - since we naturally already do it - as soon as our procedural memory of the body position for turning is established, whether seated or standing.

This is because by adding premeditated counter steering to our turning procedure, we open the door that later will allow us to learn more advanced movements like slides, drifts, and other dynamic turning techniques.

CHAPTER VII

PERCEPTION MISTAKES

OFF-ROAD IS ALSO RIDING SITTING DOWN

While I often see the world in shades of grey rather than just in black and white, I acknowledge that certain absolutes do exist, like taxes or our inevitable demise. In our dirt bike world, for many different reasons, those grey areas tend to get blurry sometimes, many times leading to dangerous misconceptions. One such prevalent misconception - often mistaken for an absolute truth - is the notion that off-road riding must invariably be done standing up, particularly on adventure bikes.

This belief, reinforced by images of riders in locked grips and poised in dynamic stances, has become a hallmark of a level of riding prowess many aspire to achieve. Such compelling imagery, frequently utilized in both personal and corporate marketing, subtly instills the idea that true off-road mastery equates to standing on the pegs and being flamboyant.

Over time, this perception has only deepened, partly due to some training programs either overly emphasizing this stance or not sufficiently exploring its alternatives, thus perpetuating this mindset among newcomers to our sport.

It's important to clarify, however, that standing up is indeed a crucial skill in off-road riding. But it's not the be-all and end-all of riding techniques, nor is it always the optimal choice for every scenario. Evidence of this lies in the practices of elite off-road athletes, including rally raid, motocross, and supercross champions, who often ride seated in various situations, showcasing that riding seating down is a valuable skill to be considered, not one to be immediately disregarded.

Now, if even the fastest off-road riders in the world, that train day in and day out to be on top of a bike don't limit themselves to standing, what leads us to believe that we, as every-other-Sunday riders on our majority, should take a different approach?

For the sake of coherency, as well as of safety when it comes to riding, I strongly believe we need to reevaluate our approach as a group, if we really wish to improve our riding consistency. With that in mind, we need to discard our preconceptions and resist the urge to immediately stand up as soon as the pavement ends.

Accepting this requires an understanding that multiple riding positions are valid and necessary, forcing us to define these positions and discerning when they are most appropriate, even if the criteria for switching between them is not easily delineated in text.

Motorcycle riding is an individual sport enjoyed collectively, and while we may ride among friends, we alone bear responsibility for our actions on the bike, including our choice of body position.

Embracing this responsibility makes it apparent that our riding group comprises individuals with varying levels of skill, experience, comfort, safety awareness, and risk management. This individuality, which makes each of us unique, underscores why it's nearly impossible to specify exact moments for position changes.

Therefore, it's essential to turn inward and develop a self-reliant decision-making process. The simple yet effective maxim I advocate, both personally and professionally, is this: maintain the least physically taxing body position until the situation demands expanding your safety envelope.

Put simply, if you're standing on terrain where riding seated down would suffice, you're expending excessive energy for the level of safety required.

Though erring on the side of caution is not inherently wrong, it can be physically draining, and given our finite physical endurance, prioritizing maximum safety at all times will inevitably lead to shorter riding sessions. Surprisingly, this often-overlooked fact explains why many leisure riders fatigue quickly, as this persistent adherence to standing can prematurely end a day's ride, eventually pushing riders beyond their physical and cognitive limits. Ironically, this misjudgment in terrain-appropriate positioning leads to such exhaustion that riders will tend to become less safe over time, as their cognitive and physical abilities decline.

The contrast in performance between a fresh and a fatigued rider is stark, highlighting the dangers of exceeding our physical limits - a cornerstone of numerous road safety campaigns. Thus, efficiently managing our energy is key to effectively controlling our personal safety envelope and determining the most suitable body position for achieving it.

THE TWO SEATED POSITIONS

Now that we've grasped the importance of energy management, we need to develop a basic understanding of what each riding position offers in terms of its relative energy expenditure.

In the curriculum I teach at my school, I like to reference four positions, ranked from least to most defensive: seated, seated attack position, standing, and standing attack position.

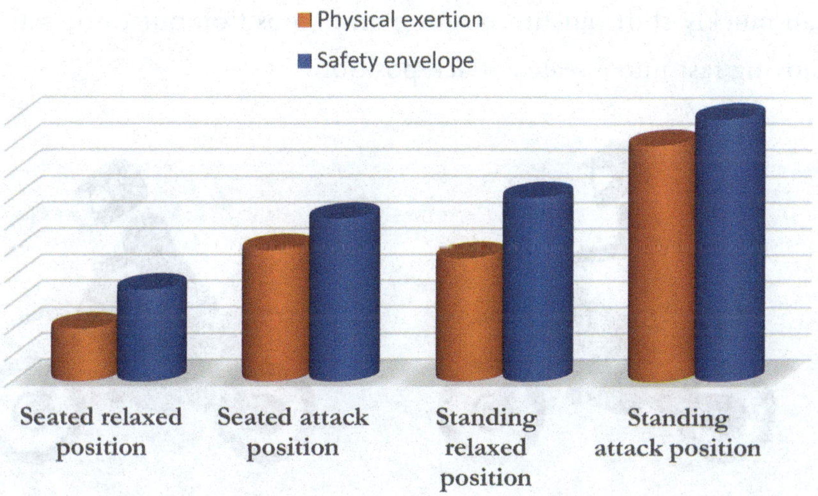

Graphic 7.1 – Comparison between safety envelope and physical exertion.

The seated position is the simplest and most intuitively understood. It's where we can relax the most. However, relaxing doesn't mean forgetting the basic position of our feet or our overall body posture; it just means allowing ourselves to ride with a relaxed hip and, if necessary, even with slightly drooping elbows and shoulders. This position is perfectly acceptable for long stretches of dirt where the terrain is regular, predictable, and where we have enough visibility to maintain higher speeds without many surprises.

For a visual reference, I would classify the type of terrain where this position is perfectly acceptable as Class 1 and Class 2, with some riders and bikes managing it quite well in Class 3, as per the terrain classification in the appendix.

That said, because we're relaxed but maintaining a good foot position, if the terrain changes, or if our confidence is shaken for any reason, we can quickly shift our hip position and reposition our torso with ease, moving fast into a seated attack position.

Image 7.1 – Difference in attitude between relaxed seated position and seated attack position.

This position, by creating separation, significantly enhances our ability to control the bike by allowing us to control the rear wheel with our body, thereby greatly increasing our safety. However, this increased safety comes at the cost of higher energy expenditure, so once we're past the challenging section that required the position change - or once we've regained our lost confidence - it's always advisable to return to our initial relaxed position, if feasible.

THE TWO STANDING POSITIONS

To fully understand the next body positions, we must keep in mind that our goal here is safety, not speed. With that said, it should be clear that when seated in an attack position, the control and aggression we can achieve are superior to that of a simple standing position in a relaxed manner. Therefore, we can often be faster while seated.

However, standing in a relaxed posture improves our field of vision compared to sitting, as well as our ability to let the bike work beneath us without influencing our body as much. There's also validity in the argument that our speed in transitioning to a standing attack position is significantly faster. In this way, by standing relaxed, we achieve these safety benefits while expending less energy than in a seated attack position.

Thus, while we may potentially lose some speed while standing relaxed, we increase our safety envelope, which leads me to recommend that, at a competitive level where speed is crucial, one should only alternate between the two attack positions - seated and standing.

However, at a leisure level, which is the focus of this book, and where endurance is key for spending many hours and miles on the bike, it's essential to use all possible positions, especially if our physical condition isn't ideal or "race ready".

With this in mind, even though we might be slower when standing, we find ourselves now in a position where we have a similar safety envelope but less physical strain, which is why this position should never be dismissed. Whether it's just to rest our legs, or to cover miles with a broader field of vision, this position can be successfully applied to various types of terrains.

At the pinnacle of our body position arsenal is the much-coveted standing attack position.

With feet perfectly positioned, heels down and toes in, knees gripping and behind the pegs, hips rotated out, elbows high, body elongated across the bike, and eyes on the horizon, we are now at the limit of what our riding technique allows.

In this position, we can be the fastest with the highest relative safety possible. However, this position is highly taxing on the body, making it virtually impossible to maintain throughout a whole day of training or trail riding, especially for the typical leisure rider who doesn't prioritize physical preparation on or off the bike.

Therefore, choosing when to use this riding position becomes crucial, and generally, looking good in photos and videos should not be a priority over our safety and endurance.

THE ELUSIVE 5TH POSITION

While I've talked about and illustrated just four riding positions, that's not entirely accurate, as there's actually an elusive fifth position we should adopt in off-road riding. It's elusive not because it's seldom used - many riders naturally adopt it - but due to the lack of emphasis it usually gets in both teaching programs, as well as in friendly tips.

This position is what I call the "hovering seat," a temporary stance where, while seated, we lift off the seat without fully standing up, holding our body in this in-between position for a brief period time.

This position is most effective when executed from an attack seated position because of the strength and stability required in the legs and torso. When used from a relaxed seated posture, we tend to exert too much force on the handlebars to support our body weight, which can result in counterproductive instability of the bike.

Nonetheless, this is a resource that can be utilized in either of the seated positions, as it allows us to prevent the bike from jolting us during small jumps or dips, creating an air cushion between our seat and the saddle.

This air cushion is why emergency braking and downhill maneuvers should be performed standing up, in order to maximize our performance. However, these two positions shouldn't be confused with the "hovering seat", regardless of how similar they look, and to some extent feel, when preformed, being that their key difference is their duration.

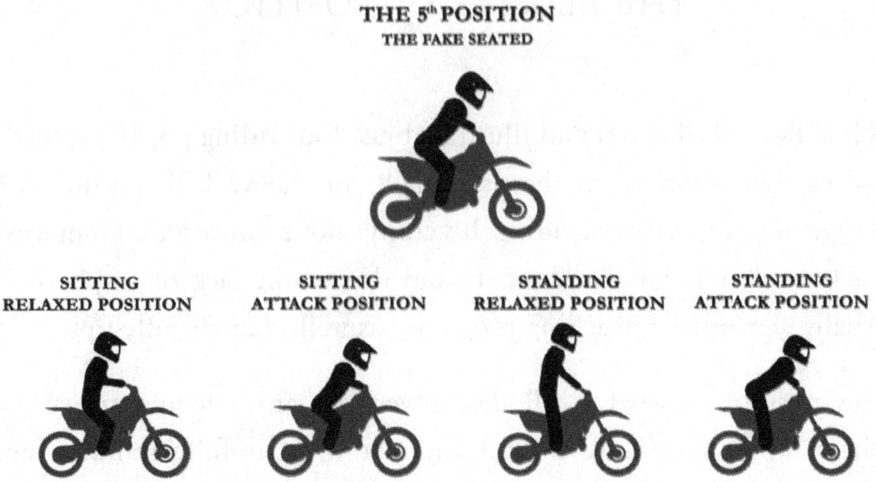

Image 7.2 – The 5 Riding Positions.

More experienced riders can also use a slight variation of this technique for pre-loading the suspension, whether to initiate a wheelie, prepare for a jump, or simply to use the bike's rebound to overcome an obstacle.

Though it's a high-energy-consuming position, since it's used momentarily and offers significant benefits, this position is an excellent choice for those wishing to maintain a high pace without the strain of continuously staying in a standing attack position or frequently transitioning between positions.

With all this in mind, it's important to remember that off-road riding is not one-dimensional. We need to train to become proficient in all riding positions and, of course, help others by sharing this information with that buddy who religiously stands up the moment they hit the dirt.

LIFTING THE BIKE OFF THE GROUND

There are various techniques for lifting motorcycles, and the typical approach of leaning your lower back - or chest - against the seat, grabbing the handlebar with one hand and a rear support with the other, then lifting from there, is relatively intuitive. However, intuitive doesn't always mean it's entirely correct, and thus, free from notes and adjustments.

That being said, in these positions, and as general notes, it's crucial to maintain a tense torso with the hips rotated out, akin to the stance found in the gym for certain weightlifting exercises. Similarly, it's important to keep in mind the lesser-discussed - and yet extremely important - notion that bikes are lifted in two stages, not in one fluid motion, a notion that I'm sure for many, is as strange as it is counterintuitive.

Except for certain positions, or for bikes with opposing cylinders or equipped with large panniers, when a bike falls, it tends to leave one or both tires in the air. This means the first stage of lifting the bike aims to ensure both tires make contact with the ground again. Typically, this first stage requires a deadlift of the bike, akin to a deadlift for those familiar with gym workouts. Once the tires touch the ground, we enter the second stage, as we're now rotating the weight of the bike around its tires, completely altering the necessary force and dynamics of the movement.

With this in mind, whether alone or with help, to prevent injuries, it's vital to sense when the tires touch the ground and adjust our posture accordingly, avoiding unnecessary and extreme impacts on the spine, usually a typical result of trying to lift the bike in one single motion.

It's also crucial to note that before starting to lift the bike, we should always analyze the situation. This analysis should consider crucial points for the maneuver's effectiveness, ensuring we have the best possible plan, akin to a snooker game where the white ball should always be played with the next shot in mind.

For instance, in a fall happens on an uphill, it might be more beneficial to rotate the bike on the ground to make it face downwards, pulling it by the front wheel to reposition it, instead of standing it up only to be unable to leave the spot where we crashed in the first place. The same can be said for falls or getting stuck in mud or sand, where changing the bike's position on the ground can be as important as getting it upright.

Another point of analysis is the bike's situation regarding the gear shifter. A bike in neutral will tend to roll with the slightest incline of the terrain as soon as the wheels touch the ground and its weight is in our hands. This can easily be resolved by ensuring a gear is engaged. However, if the bike is lying on its left side, shifting gears might not be possible. In such cases, any electrician's tape, a velcro strap - which I personally always keep tied to my handlebar - or even the velcro on some gloves, can be used to lock the front brake lever against the handlebar, ensuring that, regardless of the gear the bike was in when it fell, it won't roll when lifted.

Similarly, if the bike falls to the right side, and it's possible to select a gear to prevent the bike from rolling, deploying the kickstand is always advisable. Whether as a placebo effect or a safeguard against distraction or over-exertion, knowing that the bike will have a stopper to prevent it from falling to the side during the maneuver tends to promote a smoother and more relaxed execution of the entire movement.

Lastly, we should conclude our lifting analysis by considering the bike's weight. While on an enduro bike, we tend to carry little to no extra weight, the opposite tends to be true on an adventure bike, making it nearly impossible to lift in many situations, simple due to the added weight. So, while I believe we can and should always make a first attempt without changing our setup, if we notice our physical effort is too much or simply irrelevant for the success of the maneuver, removing panniers and extra weights can often make the difference between being able to lift a bike, or not.

Tents, food, water, or clothing, when added up, can often remove dozens of pounds from a fallen bike, and while the time to remove and replace this material may seem excessive in many cases, the energy lost around a bike that's virtually impossible to lift due to being too heavy can make the difference between being able to get back on track or not.

WHEN IN DOUBT, BRAAPP IT OUT

Every rider tends to hit certain milestones pretty quickly into their riding life. The first stall, the first ride without gear because "*it's just around the corner*," the first off-road fall, and the first time someone tells us that we need more speed to overcome this or that difficulty on the track. Step by step, the beginner's bingo card fills up quickly, and with it, some linguistic and perceptual mistakes start to embed themselves into our memory.

There's no doubt that a bit of speed can solve many off-road problems, so much so that I'd dare say it's almost like makeup for technical errors. In a world where the vast majority of brands, whether they admit it or not, heavily invest in bikes directly derived from their competition versions, it's not hard to see why, at racing pace, most bikes tend to feel easier to handle.

Potholes seem to vanish, sand becomes nothing more than dust under our wheels, jumps have soft landings, and turns seem to have grip even on the loosest surfaces. This empirical result, coupled with the notion that speed is needed, seems correlated, and that's where problems begin, because speed should not be the focal point.

Thus, it becomes clear that to use speed to our advantage, we must dispel the illusion that it promotes control, and understand its real impact on a bike. This misunderstanding is why we regularly see riders on various tracks and trails trying to be fast but, in reality, riding on what can only be defined as the "continuous threshold of a crash."

There are two power positions we should adopt on a bike: accelerating and braking. One could argue that coasting is also possible from a physical standpoint, but it's a mistake in the vast majority of situations and can hardly be considered a power position, so I'll blatantly ignore it.

Remembering that we have load transfers with acceleration and braking, and that our job as riders is to try to balance these two to maintain the bike's stability. Assuming that throttling out in case of doubt is always the best solution is inherently flawed, as it can easily force a clear destabilization of the vehicle while trying to escape an already precarious situation.

Yet, we also have what I'd call racing calculations. Let's take a jump on an MX or SX track as an example. Knowing the speed we need to attack the jump so it's neither too short nor too long is calculable. This reality illustrates well how using the maxim "when in doubt, throttle out" can smear the makeup and force a crash.

In this case, it can not only lead to landing outside the desired zone but force the application of techniques beyond our capabilities to save the situation.

I used the example of a track jump, but a sudden acceleration as technical makeup to try to cross a hole or log on a trail can, in the same way, mean losing traction and crashing head-on against the obstacle we're trying to overcome.

This means that instead of focusing on speed, we should focus on our momentum, which, as we've analyzed, in practice means good use of body technique and adjustment between the two power positions, i.e., accelerating or braking.

We can then return to our concept that riding a motorcycle should be viewed as a dance between the rider and the bike, where although both need their degree of independent freedom, if there's no harmony, someone's going to end up with stepped-on feet, or on the ground. With this analogy in mind, we should, just like in dancing, give some forewarning to our partner about what's going to happen, being proactive rather than reactive in our movements.

By being consistent and smooth in our braking and acceleration, we make our bike's reactions more predictable and its movements fluid, an attitude that increases our consistency and safety. It's important at this stage to remember that the throttle is not an on/off switch, so 1%

throttle is, by definition, accelerating, a fundamental concept especially for the fans of coasting.

In this way, knowing that accelerating in doubt is not the resource we should use, leveraging momentum is the correct term and technique. However, momentum and speed are often confused in common lexicon, and this is something we need to correct. To do so, we can, in a general and simplified way, define speed as a distance traveled in a given space of time. That allows us to define momentum as mass - which means the weight of the bike and rider - multiplied by a certain speed.

In an extremely simplified way, we can consider momentum as the amount of movement in a certain direction, or as inertia in motion, if you prefer. That said, even with the engine off, we continue to have momentum; what we don't have is constant momentum, since we're likely losing speed.

Understanding then the weight of our bike and how it moves, we begin to understand how we can move it by grasping the concept of moving mass in practical terms.

Going at 62 mph on a straight, braking hard to make a turn at 3 mph, and then accelerating as much as possible until the next hard braking becomes slower over the long run than doing everything at 31 mph, for example, if only because it is incredibly more physically exhausting to try to maintain the first option during a day of adventures, than to simply accept the second.

This tells us that having the engines screaming and trying to be super aggressive without a coherent game plan will be slower and more tiring than the attempt to be consistent and controlled all around. Therefore, the better we are technically, the more speed we can apply throughout the course, or rather, the more momentum we will be able to maintain.

All we're doing by "giving gas" to get past something or save a situation is treating ourselves like a wrecking ball, hoping that whatever we want to get past is surmountable with that approach. This makes it an extremely dangerous approach, proving that we should avoid using speed as a way to improve - or mask - technique. Thus, as previously analyzed, we should instead reduce speed as a method to exponentiate learning, improve technique, and instead think that in case of doubt, what we should do rather than giving more gas, is typically not to decrease our existing momentum.

TRAINING ISN'T EXPENSIVE

It's a common belief that off-road instruction is pricey, especially those offered by certified instructors trained as educators. While I'm not one to comment on anyone's financial situation, I still think there's a need to put the costs of training into perspective, so you don't have to believe tall tales at face value.

In a world where the concepts of expensive and cheap are defined by the value we attribute to things, clarifying the cost-benefit of two-wheeled training is vital, and that means doing some basic math.

It shouldn't be a surprise that there are various types of trainers and instructors, each with their cost, curriculum, and value they bring to the table. This reality is what causes training fees to range from just over a handful of dollars for a few hours, to several thousand dollars for multi-day courses.

My view on the value some instructors place on their time, effort, and knowledge, as well as the return on investment each student gets from choosing one training over another, is irrelevant for this discussion, as our goal here is to point out the widespread misperception about the costs of training.

With that in mind, relying on mathematical facts is the most coherent approach, so all costs presented for this explanation are average values at the time of this book, including those for gear, the bike itself, and of course, the training.

Let's then consider someone who's just entered the world of ADV - or even enduro - bought a bike for $10,000, and is contemplating a $500 training course.

If, for many, the cost of training seems high in principle, we could directly say that this expense equals only 5% of the investment made on the bike. From this perspective, we might say that 5% is a minor investment in one's active safety, but highlighting just this figure alone would be over simplifying, so we need to expand our model.

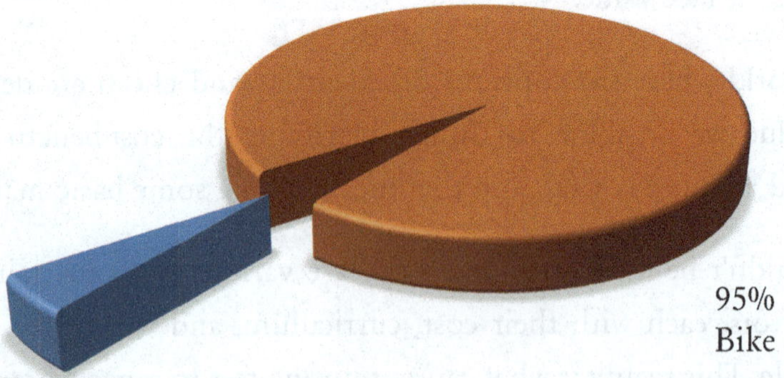

Graph 7.2 - Example of training value to be compared with the bike's price.

Let's then assume this new rider also invested $400 in a helmet, $600 in a suit with protection - or bought separate protection plus a jersey and pants - boots for $250, and gloves for $50, totaling $1,300 in protective gear. This means that this aspiring adventurer invested 13% of the bike's price of cost in its passive safety.

Speaking of protective gear, regardless of whether we consider it as passive or active safety, we must assume it is, or at least should be, replaced on average every five years of intensive use.

Let's then define "intensive use" as wearing the gear 250 days a year for this analysis. This metric will mean an investment of 13% relative to the bike's price of cost, every five years, resulting in a cost of just over $1 per day invested in our passive safety.

Conversely, the knowledge gained through training, which is applied every time we swing a leg over a bike, doesn't have a shelf life or diminished value over time. Instead, it has a constant growth through the experience gained on that training, mile after mile.

Thus, if we assume that one will ride for 20 years after buying the bike - making intensive use of it - a one-time investment of 5% in training relative to the bike's price of cost will only cost $0.07 per day over those two decades of riding.

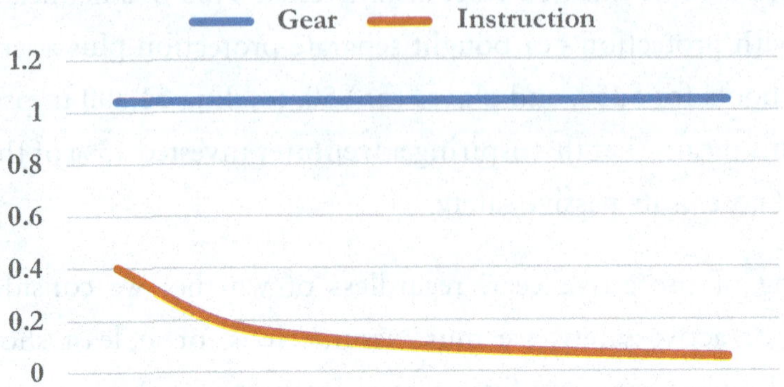

Graphic 7.3 – Best case scenario of daily cost of protective gear - over 5-year blocks - compared to the daily cost of a one-time investment in training.

This means that if we choose to take a refresher course to maintain proficiency - or even advanced courses to learn new techniques - every two years, at the same $500 per training price of cost, over those 20 years of riding, we'll invest $5,000 in our active safety. This accumulatively will cost us $0.68 per day over those two decades of adventures.

In the spirit of mathematical precision, this price of cost of $0.68 for training per day over 20 years is still less than the daily cost paid for all the protective equipment acquired in those decades, which amounts to about $0.71.

While the calculations textually demonstrate that training, despite having a "high" initial cost, actually has a nearly negligible cost over time, this is still inflated when analyzed in the real world.

This is because, on one hand, we didn't introduce the costs of extras for the bike - which almost all riders readily spend without a second thought - and on the other hand, we also didn't consider the costs of bike maintenance, both scheduled or extraordinary, regular wear and tear materials like tires or transmission kits, or even the increase in the cost of these materials over time.

Similarly, we assumed that, in 2 decades of riding, one never switched bikes, an expense that by itself would drastically alter the final results. We also assumed never needing to replace any piece of our personal protective equipment before the defined 5 years.

That said, if we applied all these metrics, the training costs would become even more diluted, especially considering that it's rare for people to take a new training course every 2 years, with most of those who choose to learn only doing 1 to 2 courses throughout their entire life.

Therefore, I have and will continue to find it hard to hear that training is expensive, whatever the price, because when there's genuinely an interest in learning how to improve our safety and performance, it's easy to see how the investment in training is not only not high, but is potentially the cheapest and highest return investment we can make in our entire two-wheeled life.

Hence, I consider and will continue to consider that this is probably the biggest perception mistake we can make as riders.

CHAPTER VIII

THE VALUE OF IGNORANCE

Discussing technical evolution in motorcycles without mentioning talent is nearly impossible. Often, this leads to some hard truths about ourselves concerning the topic. However, the negativity stems from a misunderstanding of what talent truly signifies and how we should integrate it into our overall abilities.

Talent isn't just what prodigies like Jett Lawrence, Billy Bolt, or Jeffrey Herlings display on a motorcycle. Viewing talent solely in this light is not only incorrect but also unfair to ourselves and these athletes. Talent, as defined by the dictionary, is a natural or acquired aptitude, marking the starting point from which each of us embarks on our journey in off-road riding.

Acknowledging this, we must accept that our natural talent can only take us so far; beyond that, any additional talent must be cultivated through hard work, consistency, dedication, and above all, humility and honesty towards ourselves.

The Dunning-Kruger effect, a scientific principle, illustrates how individuals with limited skills in an area tend to overestimate their abilities. This effect is likely magnified in off-road riding due to the confidence instilled by modern motorcycles. Today's bikes, with their advanced geometries, electronic aids that nearly navigate for us, and forgiving frames and suspensions, can mislead even the most self-aware riders about their true capabilities.

Conversely, older motorcycles demand a different kind of honesty, offering less margin for error, yet still more than we often acknowledge. Therefore, gauging our technical skills solely on our performance with a particular bike is a significant oversight – one I admit to having made in the past.

While a certain degree of risk and unknowns are part and parcel of our adventure-seeking spirit, caution and knowledge should guide our use of these elements. Today, I consider myself more ignorant than ever regarding motorcycle mechanics and the techniques that optimize their use. This mindset, borne from a decade of research across various scientific fields supporting our sport, fuels my continuous quest for knowledge.

This acknowledged ignorance, coupled with a disregard for my lack of natural talent as a rider, drives my relentless pursuit of learning, making me safer and more consistent than ever before. It allows me to understand my limitations and strengths better and to get the most out of my bike or of any other I ride.

Embracing my ignorance and talent deficit enables me to take pride in my limited knowledge rather than forcing me to blindly emulate others' bike modifications, chasing trends, or following paths chosen by others in hopes of matching them.

It is this acknowledgment and continuous learning from both successes and failures – thankfully still occurring – that enhance my growth the most.

Therefore, based on everything discussed from the beginning of this book, my advice is to conduct a thorough self-assessment of our theoretical knowledge and our technical abilities in the field. We should accept where we are in our evolutionary journey and make a serious decision about our real objectives.

This goal analysis should also continue during training sessions, as the purpose of any training is to identify flaws, not to glorify victories. Failures and mistakes teach us much more about our limits and true capabilities than a merely surpassable technical maneuver, which in those terms is often executed with dubious success.

Not all of us aspire to be athletes competing at the highest level, exceptionally fast riders, or show-offs. Many of us, I would venture to say nearly a majority, just want to get our bikes dirty, have fun for a few hours, and return home in one piece to share love with our families until we get the chance to coat our bikes with a few more layers of dirt.

In this regard, embracing our ignorance will be crucial for our success, as it will be our own, unique, and private way to define our theoretical and practical abilities, the safety margins we want, and devise the best plan to never exceed them. Whether it's in terms of choosing bikes, terrains, speeds, or attitudes on the trails.

So, it's time for us, as a community, to put aside our ego, keep our bravado well tucked away from the throttle, and proudly embrace the ignorance and lack of talent which, at the end of the day, is what will help us evolve and stay alive and safe every time we turn the key on.

ANNEX I

RIDER AND TERRAIN CLASSIFICATION SYSTEM

By far, one of the most challenging aspects of off-road riding isn't the riding itself but rather breaking the language barrier that exists among off-road enthusiasts, even when speaking the same base language. What's difficult terrain for many can be easy for others, and this dichotomy is enough to create serious misunderstandings.

Therefore, it becomes essential for me as a professional instructor to aim at more than just helping to develop off-road riding techniques or showcasing landscapes and tracks known for their beauty. It is equally crucial to try to eliminate points of discord and division in our sport. With this in mind, I have adopted the terrain and adventurer classification system developed by Bret Tkacs of Puget Sound Safety Off-Road (PSSOR).

All my training follows this classification.

It's important to note that some considerations and assumptions need to be taken into account to use this system, and a healthy dose of honesty and introspection is advised when applying it. To facilitate its understanding and application, it has been divided into three sections:

- **CONSIDERANTIONS AND ASSUMPTIONS**
- **ADVENTURER CLASSIFICATION**
- **TERRAIN CLASSES**

CONSIDERANTIONS AND ASSUMPTIONS

GENERAL CONSIDERATIONS FOR TERRAIN CLASSES

MOTORCYCLE - You are riding a medium to high-displacement adventure motorcycle (650cc or higher).

MODS - Your ADV motorcycle is close to its original form, with only minor alterations for protection, comfort, or luggage. Small changes like suspension tuning (for weight adjustment, for example) or improvements in brakes or tires do not count as altering the motorcycle from its original state for the purpose of this terrain classification.

USAGE - You are traveling or camping, and therefore, your motorcycle will be equipped with a luggage system filled to at least 50%.

CONSIDERATIONS FOR ADVENTURER CLASSIFICATIONS

TRANSITION - Refers to the period of time when you are developing the techniques that will allow you to reach the proficient level.

Surviving or managing to overcome certain challenges on the tracks is not a sign of proficiency.

SPEED - There is no direct connection between speed and technique. The more experienced you become, the faster you will be, but that doesn't mean you choose to do it systematically.

Different people have different comfort zones for various reasons, from personal preferences from a more general point of view to an injury or a motorcycle with limited capability from a more detailed perspective.

ADV is not trial or competition, so speed is not synonymous with technical ability.

CONSIDER YOU ARE IN A BEGINNER LEVEL IF:

STANDING UP - You don't ride standing up on moderate (or higher) difficulty terrains and prefer to sit on technical terrains.

FATIGUE - You feel tense when riding off-road and find off-road riding physically exhausting.

SPEED CONTROL - You use the throttle to control your speed and not the clutch (or the rear brake for motorcycles with DCT).

BRAKING - You don't use the front brake off-road or don't feel comfortable braking off-road.

FEET - You try to "paddle" on certain types of terrain and prefer to put your feet down on gravel, sand, or loose rocks.

INFOGRAPHIC:

Graph Annex I.1 – Terrain classification infographic.

ADVENTURER CLASSIFICATION SYSTEM

Begin by determining your natural state by assessing your well-being after a road motorcycle ride. Identify stress and fatigue levels, your ability to multitask (e.g. ride, listen to music, and adjust the GPS at the same time) if necessary, and how many breaks you typically need to take.

Now, compare your natural state with each terrain class and determine your level in each one. For example, if riding on moderate terrains (Class 3) leaves you exhausted - compared to your natural state on the road - and you regularly fall, then you are at a beginner level on moderate terrains.

BEGINNER

Consider yourself a beginner if two or more of the following situations apply to you:

- Falls or near-falls are frequent.
- Accidents or near-accidents occur frequently.
- You are exhausted, or you continually need to use a lot of energy to ride.
- You expect (or it regularly happens) to have damage to the motorcycle due to the type of terrain (pinch flats, damaged rims, extreme damage to the skid plate or engine protection, etc.).
- You are unable to multitask while riding if needed.

TRANSITION

Consider yourself transitioning to proficient if:

- Dropping the motorcycle, falls, or near-falls are infrequent.
- You are not damaging the motorcycle due to the terrain.
- You are not getting tired because of the type of terrain.
- You take breaks to eat but not due to physical fatigue.
- You can multitask while riding.

PROFICIENT

Consider yourself proficient if:

- You don't fall or have near-falls.
- You don't anticipate any type of damage to the motorcycle due to the terrain.
- You use very little energy to ride.
- Your breaks are for eating, due to weather conditions, or for stretching, but not because of physical fatigue.
- You can easily multitask while riding, if necessary.

TERRAIN CLASSES

CLASS 1 / BEGINNER TERRAIN ⊖

Consider yourself fit as a beginner if you are proficient in this class of terrain. Consider it basic training if you're struggling with this class of terrain or if you're seeking continuous training from scratch.

CONCEPT

The beginner class is where most of us start, even before considering becoming an ADV Rider. Gravel roads, well-maintained or packed dirt roads mark the beginning of this class. This type of terrain and features can be navigated with road motorcycles equipped with street tires and doesn't necessarily require specific off-road training.

CHARACTERISTICS

- Old or unmaintained asphalt roads
- Gravel roads or well-kept dirt roads
- Small sticks and branches
- Water crossings less than 5 cm or 2" deep
- Modest inclines and descents (similar to those found on paved roads)

CLASS 2 / BASIC TERRAIN ●

Consider yourself fit for the basic level if you are proficient in this class of terrain.

Consider basic training to learn how to handle the challenges of this class and start your continuous training.

CONCEPT

Basic terrains are the most common for ADV Riders of all levels.

This type of terrain can be crossed by a regular car.

Basic terrains consist mostly of simple, predictable features, with the occasional traction challenge or obstacles that require slowing down before crossing.

CHARACTERISTICS

- Wide single-track or wide two-way roads with a central divider
- Wide and shallow ruts
- Packed dirt
- Washboard-like wavy surfaces
- Obstacles or edges less than 10 cm or 4" in height
- Sections of loose rocks less than 8 cm or 3" deep
- Sections of loose gravel, shallow sand, or shallow mud
- Water crossings with weak current and less than 10 cm or 4" deep
- Water crossings with a solid base
- Small rocks may roll under the tire
- Shallow gravel with rain grooves

CLASS 3 / MODERATE TERRAIN ■

Consider yourself fit for the moderate (or intermediate) level if you are proficient in this class of terrain.

Consider advanced training to learn to handle the difficulties of this class.

CONCEPT

Moderate terrains are where good off-road training and technique start to show their importance.

This is the type of terrain most sought after by ADV Riders when traveling. Striving for proficiency at this level is an excellent goal to make the most of your bike.

It's in this type of terrain where skills like riding well standing up and having precise clutch control are necessary to become proficient.

CHARACTERISTICS

- Single-track with ruts (dry terrain)
- Sections of shallow mud
- Loose shallow gravel (gravel from road maintenance)
- Sections of loose sand less than 30m or 100ft long
- Wet grass
- Obstacles up to 15 cm or 6" in height
- Water crossings with mud or loose rocks as a base
- Water crossings less than 15 cm or 6" deep

- Sections of wide zigzagging single-track
- Sections of loose rocks less than 13 cm or 5" in diameter on inclines
- Steps less than 10 cm or 4" in uphill climbs
- Steps less than 15 cm or 6" in descents

CLASS 4 / HARD TERRAIN ◆

Consider yourself fit for the difficult (or advanced) level if you are proficient in this class of terrain.

Consider advanced training or private lessons to learn how to handle the difficulties of this class.

CONCEPT

For many, this is where the adventure fun begins, with most tracks having only a few small sections of hard terrain. This means that this class of terrain may not be problematic for someone proficient at the intermediate level. However, if a significant portion of the track is classified as difficult, it can contribute to significant fatigue that will lead to technical and/or judgment errors in less prepared riders.

Hard terrains can be quite dangerous for those who don't have the right safety equipment (including motocross boots), the correct technique, or a properly tuned bike. Hard terrains may require high ground clearance or 4x4 vehicles to be passable.

CHARACTERISTICS

- Narrow two-way sections with a central divider in zigzags
- Snow
- Large loose rocks larger than 13 cm or 5" in diameter
- Single-track with ruts (wet surface)
- Deep and loose gravel over 8 cm or 3" deep
- Mud or sand that may require a push to get through
- Water crossings without a solid base
- Water crossings with moderate current less than 23 cm or 9" deep
- Water crossings with strong current less than 18 cm or 7" deep
- Loose rocks less than 15 cm or 6" in diameter
- Mud or water in uphill sections

CLASS 5 / EXTREME TERRAIN ❖

Consider yourself fit for the extreme (or professional) level if you are proficient in this class of terrain.

Consider individual or advanced and specific training to learn how to handle the difficulties of this class.

CONCEPT

Since this classification system assumes that you are riding a loaded ADV bike, it is not common to encounter this type of terrain by accident.

Many may encounter this type of terrain on long journeys, but few become proficient.

A professional in this class can navigate these paths in a relaxed, comfortable, and controlled manner, although they are more suitable for enduro bikes.

CHARACTERISTICS

- Trails designed for enduro bikes
- Sections of zigzagging single-track
- Impassable unless conditions are perfect
- May require mechanical assistance (winch)
- Narrow sections with lateral camber
- Loose gravel over 8 cm or 3" deep
- Mud or sand that may require a push to get through
- Water crossings with strong currents and more than 30 cm or 12" deep
- Water crossings with a loose base
- Obstacles or edges equal to or higher than the front wheel axle
- Deep and loose sand
- Viscous mud

ANNEX II

FIRST AID CHEAT SHEET

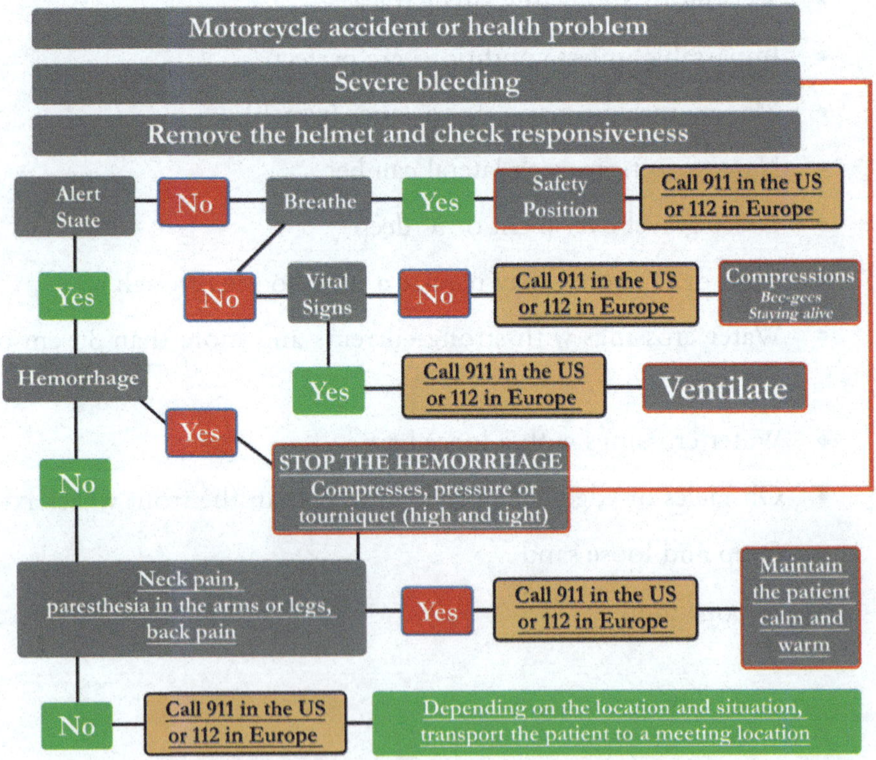

Graph Annex II.1 – First aid sheet cheat.

REFERENCES

ADVMotoSkillZ. (2020, October 11). *Home* [YouTube channel]. YouTube. Retrieved May 4, 2023, from https://www.youtube.com/@advmotoskillz/

Andrew Huberman. (2022, August 15). *Dr. Peter Attia: Exercise, Nutrition, Hormones for Vitality & Longevity | Huberman Lab Podcast #85* [Video]. YouTube. https://www.youtube.com/watch?v=DTCmprPCDqc

Andrew Huberman. (2023, April 24). *Science-Based Mental Training & Visualization for Improved Learning | Huberman Lab Podcast* [Video]. YouTube. https://www.youtube.com/watch?v=0RYyQRQFgFk

Andrew Huberman. (2023, February 20). *How to Breathe Correctly for Optimal Health, Mood, Learning & Performance | Huberman Lab Podcast* [Video]. YouTube. https://www.youtube.com/watch?v=x4m_PdFbu-s

Andrew Huberman. (2023, March 6). *How to Optimize Your Water Quality & Intake for Health | Huberman Lab Podcast* [Video]. YouTube. https://www.youtube.com/watch?v=at37Y8rKDlA

Brake Magazine. (2014, April 10). *Home* [YouTube channel]. YouTube. Retrieved May 4, 2023, from https://www.youtube.com/@BrakeMagazine

Bret Tkacs. (2007, September 28). *Home* [YouTube channel]. YouTube. Retrieved May 4, 2023, from https://www.youtube.com/@BretTkacs

Bret Tkacs. (2022, October 6). *ADV Skill Rating.* https://brettkacs.com/adv-skill-rating-system/

Chris Birch - Off Road Coach. (2015, July 20). *Hom*e [YouTube channel]. YouTube. Retrieved May 4, 2023, from https://www.youtube.com/@chrisbirch-offroadcoach9177

Coach Robb Beams. (2019, January 29). *Coach Robb-How to Maintain Proper Hydration.* [Video]. YouTube. https://youtu.be/9nbhNMog28w

Coach Robb Beams. (2022, June 1). *Coach Robb: Coach Robb Podcast #55 - Spasms and Cramps* [Video]. YouTube. https://www.youtube.com/watch?v=zcXt_MJ..gAxA

Cossalter, V. (2006). Motorcycle Dynamics. Lulu.com.

Engineering Explained. (2011, September 30). *Traction Circle - Explained* [Video]. YouTube. https://www.youtube.com/watch?v=JjCcFsGLpaM

Engineering Explained. (2015, April 22). *Deriving Car Slip Angles - Tire Slip - WARNING: Maths* [Video]. YouTube. https://www.youtube.com/watch?v=1NTdXhLnA_w

Engineering Explained. (2015, April 8). *Slip Angles - Tire Traction - Explained* [Video]. YouTube. https://www.youtube.com/watch?v=9bs2cEyK7Uo

Foale, T. (2006). Motorcycle Handling and Chassis Design: The Art and Science. Tony Foale.

FortNine. (2021, September 12). *Motorcycle Riders - You're Leaning the Wrong Way* [Video]. YouTube. https://www.youtube.com/watch?v=U1mSavQ_DXs

FortNine. (2022, August 13). *Motorcycle Riders – Your Throttle Grip is Wrong* [Video]. YouTube.

https://www.youtube.com/watch?v=tPsjr7jxO0o

GYPSY TALES PODCAST. (2021, December 30). *CHAPTER 198 Ft. Chase Sexton - Gypsy Tales Podcast* [Video]. YouTube. https://www.youtube.com/watch?v=-2EkNVdNU1s

GYPSY TALES PODCAST. (2021, July 27). *CHAPTER 173 Ft Todd Waters* [Video]. YouTube. https://www.youtube.com/watch?v=o4mzUIU7PR4

GYPSY TALES PODCAST. (2021, July 27). *CHAPTER 173 Ft Todd Waters* [Video]. YouTube. https://www.youtube.com/watch?v=o4mzUIU7PR4

GYPSY TALES PODCAST. (2021, June 13). *Gypsy Tales - CHAPTER 153 Ft. Ryan Hughes - FULL PODCAST* [Video]. YouTube. https://www.youtube.com/watch?v=bl__urhYgJM

GYPSY TALES PODCAST. (2021, September 26). *CHAPTER 184 Ft. Jordan Sullivan - Gypsy Tales Podcast* [Video]. YouTube. https://www.youtube.com/watch?v=gvVNvk6YiKM

GYPSY TALES PODCAST. (2022, December 15). *CHAPTER 147 ft. AJ Catanzaro - Gypsy Tales Podcast* [Video]. YouTube. https://www.youtube.com/watch?v=tFpEB-Q6WqY

GYPSY TALES PODCAST. (2022, January 31). *CHAPTER 203 Ft Casey Stoner - Gypsy Tales Podcast* [Video]. YouTube. https://www.youtube.com/watch?v=cbIq2jymnxw

GYPSY TALES PODCAST. (2022, June 13). *CHAPTER 219 Ft. Dr Steve Andrews - Gypsy Tales Podcast* [Video]. YouTube. https://www.youtube.com/watch?v=xkmnXRk1DLo

GYPSY TALES. (2019, January 14). *CHAPTER XLVI Ft. Robb Beams - Elite Performance Coach* [Video]. YouTube. https://www.youtube.com/watch?v=_HVdGqTQ26Q

GYPSY TALES. (2020, April 29). *CHAPTER XCIX Ft Todd Waters &*

Robb Beams [Video]. YouTube.
https://www.youtube.com/watch?v=WSOLqUCFZ3U

Gypsy Tales. (2021, August 9). *What is the relationship between TECHNIQUE and TALENT? - A conversation about Motocross Technique.* [Video]. YouTube. https://youtu.be/g1zc4kH7uYM

GYPSY TALES. (2021, December 11). *Chase Sexton explains his flawless technique on a dirtbike - Gypsy Tales Podcast* [Video]. YouTube.
https://www.youtube.com/watch?v=zpRYLCos1Oo

Gypsy Tales. (2021, September 23). *How to hydrate for racing Motocross - PHD Dietetics Jordan Sullivan explains - Gypsy Tales.* [Video]. YouTube.
https://youtu.be/ot6TpyD4BoI?t=61

Haney, P. (2003). The Racing & High-performance Tire: Using the Tires to Tune for Grip and Balance. Society of Automotive Engineers.

Jordan B Peterson. (2022, October 13). *Neuroscience Meets Psychology | Dr. Andrew Huberman | EP 296* [Video]. YouTube.
https://www.youtube.com/watch?v=z-mJEZbHFLs

Lesics. (2019, December 30). *Counter Steering | The interesting physics behind it.* [Video]. YouTube. https://youtu.be/ZpV2Bg-WX0w

Lesics. (2020, September 29). *Understanding Rolling Resistance!.* [Video]. YouTube. https://youtu.be/_S2lyaMgBQ8

Lesics. (2021, January 30). *The importance of tire slip*. [Video]. YouTube.
https://youtu.be/tsnYqCRWTbE

Mike on Bikes. (2020, July 6). *Tire Grip | EXPLAINED* [Video]. YouTube. https://www.youtube.com/watch?v=vl9gJ1aQg1E

Mike on Bikes. (2021, August 9). *How Motorcycles Corner | EXPLAINED.* [Video]. YouTube. https://youtu.be/5s_JCuByuEI

MOTOTREK. (2017, March 10). *Home* [YouTube channel]. YouTube. Retrieved May 4, 2023, from https://www.youtube.com/@MOTOTREK

Peter Attia MD. (2022, March 21). *Hydration — electrolytes, supplements, sports drinks, & performance effects [AMA 33 Sneak Peek]* [Video]. YouTube. https://www.youtube.com/watch?v=tcMfBS3BMjY

Physiotutors. (2018, January 13). *MUSCLE CRAMPS EXPLAINED by Science* [Video]. YouTube. https://www.youtube.com/watch?v=ayql0E0-L3E

Ryan Hughes. (2020, May 24). *Home* [YouTube channel]. YouTube. Retrieved May 4, 2023, from https://www.youtube.com/@RyanHughesMX

Ryan Hughes. (2020, September 1). *Hips Out - The Most Important Tip in Motocross from Ryan Hughes.* [Video]. YouTube. https://www.youtube.com/watch?v=kcIu0-7aCt0

Ryan Hughes. (2021, May 14). *Are You a Reactor or Initiator? - Motocross Technique* [Video]. YouTube. https://www.youtube.com/watch?v=FcogaKtKeEI

Ryan Hughes. (2022, January 29). *Understanding the mind when riding.* [Video]. YouTube. https://youtu.be/qIvyidYmbg4

Schwellnus, M. (2009). Cause of exercise-associated muscle cramps (EAMC)–altered neuromuscular control, dehydration, or electrolyte depletion? British Journal of Sports Medicine, 43, 401-408.

TEDx Talks. (2014, August 5). *Learning how to learn | Barbara Oakley | TEDxOaklandUniversity* [Video]. YouTube. https://www.youtube.com/watch?v=O96fE1E-rf8

TEDx Talks. (2019, June 28). *What Motorcycling Taught Me About Dealing with Anxiety | Benjamin White | TEDxBlinnCollege* [Video]. YouTube. https://www.youtube.com/watch?v=Pi5WZdW6GQY

The Moto Academy Podcast Clips. (2022, November 7). *Jett gives lower body riding tips!!* [Video]. YouTube.

https://www.youtube.com/watch?v=DQenp61isYA

The Moto Academy. (n.d.). The Moto Academy. https://club.themotoacademy.com/

Tony Foale Designs, article on motorcycle tire characteristics and how they work. (n.d.). https://motochassis.com/Articles/Tyres/TYRES.htm

Tözeren, A. (2006). Human Body Dynamics: Classical Mechanics and Human Movement. Springer Science & Business Media.

Made in United States
North Haven, CT
10 November 2026

22096533R00117